F

The Practical Approach Series

SERIES EDITORS
D. RICKWOOD
Department of Biology, University of Essex
Wivenhoe Park, Colchester, Essex CO4 3SQ, UK

B. D. HAMES
Department of Biochemistry, University of Leeds
Leeds LS2 9JT, UK

Affinity chromatography
Animal cell culture
Animal virus pathogenesis
Antibodies I and II
Biochemical toxicology
Biological membranes
Biosensors
Carbohydrate analysis
Cell growth and division
Centrifugation (2nd edition)
Clinical immunology
Computers in microbiology
DNA cloning I, II, and III
Drosophila
Electron microscopy in molecular biology
Fermentation
Flow cytometry
Gel electrophoresis of nucleic acids (2nd edition)
Gel electrophoresis of proteins (2nd edition)
Genome analysis
HPLC of small molecules
HPLC of macromolecules
Human cytogenetics
Human genetic diseases
Immobilised cells and enzymes
Iodinated density gradient media
Light microscopy in biology
Liposomes
Lymphocytes
Lymphokines and interferons
Mammalian development
Medical bacteriology
Medical mycology
Microcomputers in biology
Microcomputers in physiology
Mitochondria
Mutagenicity testing
Neurochemistry
Nucleic acid and protein sequence analysis
Nucleic acids hybridisation
Nucleic acids sequencing
Oligonucleotide synthesis
Peptide hormone action
Peptide hormone secretion
Photosynthesis: energy transduction
Plant cell culture
Plant molecular biology
Plasmids
Post-implantation mammalian development
Prostaglandins and related substances
Protein function
Protein purification applications
Protein purification methods
Protein sequencing
Protein structure
Proteolytic enzymes
Radioisotopes in biology
Receptor biochemistry
Receptor – effector coupling
Ribosomes and protein synthesis
Solid phase peptide synthesis
Spectrophotometry and spectrofluorimetry
Steroid hormones
Teratocarcinomas and embryonic stem cells
Transcription and translation
Virology
Yeast

Biosensors
A Practical Approach

Edited by
A. E. G. CASS

*Imperial College of Science and Technology,
Centre for Biotechnology, London SW7, UK*

Oxford University Press, Walton Street, Oxford OX2 6DP

Oxford is a trade mark of Oxford University Press

*Published in the United States
by Oxford University Press, New York*

© *Oxford University Press 1990*

*All rights reserved. No part of this publication may be reproduced,
stored in a retrieval system, or transmitted, in any form or by any means,
electronic, mechanical, photocopying, recording, or otherwise, without
the prior permission of Oxford University Press*

*This book is sold subject to the condition that it shall not, by way
of trade or otherwise, be lent, re-sold, hired out or otherwise circulated
without the publisher's prior consent in any form of binding or cover
other than that in which it is published and without a similar condition
including this condition being imposed on the subsequent purchaser*

*British Library Cataloguing in Publication Data
Biosensors. 1. Biosensors I. Cass, A. E. G. 547
ISBN 0-19-963046-1
ISBN-0-19-963047-X (pbk.)*

*Library of Congress Cataloging in Publication Data
Biosensors: a practical approach/edited by A. E. G. Cass.
(Practical approach series)
Includes bibliographical references.
1. Biosensors. I. Cass, A. E. G. II. Series.
R857.B54B55 1989   610'.28—dc20   89-23012
ISBN 0-19-963046-1
ISBN 0-19-963047-X (pbk.)*

Typeset and printed by Information Press Ltd, Eynsham, Oxford

Preface

THE earliest recorded applications of enzymes in analytical chemistry were in the 1860s when methods for the determination of starch using a malt extract and hydrogen peroxide using a malt extract and guaiac tincture were described. Since then the power of enzymatic methods has been extensively developed and other biological materials have been recruited by analytical science.

In parallel with this increasing use of biological methods there has been an equal if not greater growth in the variety and sophistication of analytical instrumentation with more and more physical phenomena being exploited in the identification and quantitation of substances. A third strand in the modern analytical triad is the increasing use of computer-based methods to control and process analytical equipment and data. In particular the disciplines of chemometrics and expert systems are starting to have an impact on the collection and reporting of analytical results.

Until quite recently the 'wet' biological methods and the 'dry' instrumentation have tended to develop separately and it is the increasingly close coupling of these that forms the topic of this book. The advantages of directly transducing a biological reaction into an electrical signal are several-fold and include high specificity, ease of use and the opportunity to base analysis of substances on their biological action rather than their molecular structure.

Although many combinations of biological material and transducer have been developed as biosensors the examples in this book are based largely upon electrochemical devices. These have the advantage of being applicable to many different biological systems offering many different analyses using the same basic principle. Furthermore the resources needed to construct electrochemical sensors are relatively non-specialized and the principles of their operation relatively straightforward. I hope that this will encourage readers with a biological sciences background to take them up. At the same time many of the enzymes used in these sensors are commercially available and therefore I also hope that chemists and physicists will be encouraged by what has already been achieved to develop new and sophisticated methods for measuring biological activity.

Some applications of electrochemical methods have already been described in a companion title in this series (*Immobilised cells and enzymes: a practical approach*, edited by J.Woodward) and readers are encouraged to consult this source as well as the present volume.

A. E. G. CASS

Contributors

PHILIP N. BARTLETT
Department of Chemistry,
University of Warwick, Coventry, Warks CV4 7AL, UK.

BENGT DANIELSSON
Pure and Applied Biochemistry,
The Chemical Centre, Box 124,
University of Lund, S-22100 Lund, Sweden.

CHRISTOPHER L. DAVEY
Department of Biological Sciences,
University College of Wales,
Aberystwyth, Dyfed SY23 3DA, UK.

MARK EDDOWES
Thorn EMI Central Research Laboratories,
Dawley Road, Hayes, Middlesex UB3 1HH, UK.

NICOLA C. FOULDS
MediSense (UK) Inc.,
14 Blacklands Way,
Abingdon, Oxon OX14 1DY, UK.

JANE E. FREW
MediSense (UK) Inc.,
14 Blacklands Way,
Abingdon, Oxon OX14 1DY, UK.

MONIKA J. GREEN
MediSense (UK) Inc.,
14 Blacklands Way,
Abingdon, Oxon OX14 1DY, UK.

H. ALLEN O. HILL
Inorganic Chemistry Laboratory,
University of Oxford, South Parks Road, Oxford OX1 3QR, UK.

ISAO KARUBE
Research Centre for Advanced Science and Technology,
University of Tokyo, Komaba, Meguro-ku, Tokyo 153, Japan.

DOUGLAS B. KELL
Department of Biological Sciences,
University College of Wales,
Aberystwyth, Dyfed SY23 3DA, UK.

GURDIAL S. SANGHERA

Inorganic Chemistry Laboratory,
University of Oxford, South Parks Road, Oxford OX1 3QR, UK.

MASAYUKI SUZUKI

Research Centre for Advanced Science and Technology,
University of Tokyo, Komaba, Meguro-ko, Tokyo, **153**, Japan.

DANIEL THEVENOT

Laboratorie de Bioelectrochimie et d'Analyse du Milieu,
Université Paris Val de Marne, 94010 Créteil Cedex, France.

GEORGE S. WILSON

Department of Chemistry,
University of Kansas, Lawrence, KS 66045-0046, USA.

FREDRIK WINQUIST

Laboratory of Applied Physics,
Department of Physics and Measurement Technology,
Linköping University, S-58183 Linköping, Sweden.

Contents

Abbreviations	xv

1 Unmediated amperometric enzyme electrodes 1
George S. Wilson and Daniel R. Thévenot

1. Introduction	1
2. Basic techniques	3
Enzyme immobilization	3
Protective membranes	7
Cell and sensor geometry	10
Instrumentation	12
Characterization of sensor response	14
3. Conclusions	16
Acknowledgements	16
References	16

2 Mediated amperometric enzyme electrodes 19
H. Allen O. Hill and Gurdial S. Sanghera

1. Aim	19
Introduction	19
2. Cyclic voltammetry	20
Catalytic systems	23
3. Ferrocene-based glucose sensor	23
Experimental methods	24
Analysis and optimization of performance	26
4. Creatine kinase assay based on ferrocene	30
An ATP biosensor	30
Estimation of creatine kinase activity	31
5. Ferrocene-mediated cholesterol biosensor	32
Dehydrogenase biosensor	32
Oxidase biosensor	32
Peroxidase/oxidase biosensor	34
6. Electrochemistry of redox proteins	35
Modified gold electrodes	35
Graphite electrodes	37
Electrochemically driven respiration in mitochondria	38
A lactate biosensor	41

	Electrosynthesis of p-hydroxybenzaldehyde	43
	Horizons	44
	Acknowledgements	45
	References	45

3 Conducting organic salt electrodes 47
Philip N. Bartlett

1. Introduction 47
    Conducting organic salts 47

2. Preparation of conducting organic salts 49
    Preparation of TTF.TCNQ 50
    Preparation of NMP.TCNQ 51
    Preparation of single crystals 53
    Other conducting organic salts 55

3. Electrochemistry 55
    Instrumentation 55
    Electrodes 58
    Electrochemical cells and solutions 62

4. Preparation of organic conducting salt electrodes 64
    Drop coated electrodes 64
    Packed cavity electrodes 66
    Paste electrodes 67
    Pressed pellet electrodes 68
    Single crystal electrodes 68

5. Electrochemistry of conducting organic salts 70
    Electrochemistry of TTF.TCNQ 70
    Electrochemistry of NMP.TCNQ 73
    Other conducting organic salts 75

6. Enzyme electrodes 75
    Flavoproteins 75
    NADH-dependent dehydrogenases 82

7. The analysis of enzyme electrode data 87
    The model 87
    Analysis of results 89

Acknowledgements 94
References 94

4 Immunoelectrodes 97
Nicola C. Foulds, Jane E. Frew, and Monika J. Green

1. Introduction 97

2. Basic concepts of immunoassays 97
    The structure of antibodies 97
    Classification of immunoassays 99

Enzymes as labels in immunoassays	100
3. Electrochemical techniques for the development of amperometric immunoassays	100
DC cyclic voltammetry	101
Chronoamperometry	103
4. Alkaline phosphatase-labelled electrochemical immunoassays	103
Properties of alkaline phosphatases	103
Measurement of alkaline phosphatase activity	104
Alkaline phosphatase as an enzyme label	105
Electrochemically active substrates for alkaline phosphatase	105
Hapten conjugation	106
Antibody purification	111
Solid-phase immobilization of antibodies	112
Configuration of electrochemical immunoassays	114
5. Glucose oxidase in electrochemical immunoassays	116
Properties of glucose oxidase	116
Measurement of glucose oxidase activity	116
Electrochemical detection of glucose oxidase	118
Preparation of enzyme conjugates	119
Preparation of hapten–mediator conjugates	121
Configuration of electrochemical immunoassays	121
References	124

5 Conductimetric and impedimetric devices   125
Douglas B. Kell and Christopher L. Davey

1. Introduction	125
2. Theory of the impedimetric experiment	125
The concepts of impedance and admittance	125
Intrinsic system properties and dielectric relaxation	128
Mechanisms of dielectric relaxation in biological systems	133
3. Electrodes	134
Polarizable and non-polarizable electrodes	134
The impedance of elemental electrodes	136
4-Terminal systems	140
4. Hardware requirements	140
Frequency-domain methods; analogue	141
Digital frequency and time-domain methods	142
Other measurement considerations	144
5. Implementation and commercial devices	145
Enzymatic	145
Microorganisms	145
Tissues	151
6. Concluding remarks	152

	Acknowledgements	152
	References	152

6 Microbial biosensors   155
Isao Karube and Masayasu Suzuki

1.	Introduction	155
2.	Principles of microbial sensors	155
3.	Construction of microbial sensors	157
	Immobilization of microorganisms	157
	Electrochemical devices for microbial biosensors	159
	Construction of microbial biosensors	161
4.	Applications of microbial biosensors	165
	Application fields of microbial biosensors	165
	BOD sensor	165
	Gas sensors	166
	Electrochemical bioassay	167
	Hybrid biosensor	168
	References	169

7 Semiconductor field effect devices   171
Fredrik Winquist and Bengt Danielsson

1.	Introduction	171
2.	Basic principles of field effect devices	172
	A brief survey of semiconductor physics	172
	Ion selective field effect transistors—ISFETs	173
	Gas sensitive metal oxide semiconductor structures	174
	Principal differences between ISFETs and gas sensitive MOS devices	176
3.	Experimental	176
	Sensor fabrication and instrumentation	176
	Preparation of enzymes	179
	Immobilization of enzymes	180
	Biosensing systems	182
4.	Methods	184
	Bioanalytical uses of pH sensitive ISFETs	184
	PdMOSFET in bioanalysis	185
	Ammonia sensitive IrTMOS in bioanalysis	186
5.	Summary	188
	Acknowledgements	190
	References	190

8 Thermometric sensors   191
Bengt Danielsson and Fredrik Winquist

1.	Introduction	191
2.	Different principal designs of thermal biosensors	192
3.	Instrumentation	192
	A simple plexiglas apparatus	192
	Present enzyme thermistor design with aluminium calorimeter	194
	A miniaturized enzyme thermistor	195
4.	Procedures	196
5.	Applications	197
	Applications based on immobilized enzymes	198
	Applications based on immobilized cells	201
	Chemical and enzymic amplification	203
	Enzyme activity determinations	204
	Measurements on cell suspensions	205
	Process monitoring and control	206
	Environmental control applications	206
	TELISA (Thermometric enzyme linked immunosorbent assay)	207
6.	Concluding remarks	208
	References	208

9 Theoretical methods for analysing biosensor performance 211

Mark J. Eddowes

1.	Introduction	211
	Role of theory	211
	Fundamental aspects of biosensor operation	211
	Mathematical tools	212
2.	Biological reactions	212
	Homogeneous single site binding equilibrium	213
	Binding at surfaces: the adsorption isotherm	215
	Heterogeneity of binding sites, multivalence and cooperativity	216
	Enzyme kinetics	216
3.	Mass transport	219
	Diffusion as a random process	219
	Fick's laws of diffusion	220
	Application of Fick's laws to diffusion problems	221
	Fundamentals of convective mass transport	223
	Transport due to migration	226
4.	Coupled transport and reaction processes: analysis of complete systems	227
	Steady-state analysis of mediated amperometric systems	228
	Immobilized enzyme layers with non-consuming reaction product detection	233
	Response time of a diffusion membrane: solution of a time-dependent problem	243

5. Mathematical methods	247
Analytical solution	248
Numerical solution	258
References	262
Appendix Suppliers of reagents, membranes, sensors and associated equipment	265
Index	269

Abbreviations

BSA	bovine serum albumin
CK	creatine kinase
COD	cholesterol oxidase
CPG	controlled pore glass
CVD	chemical vapour deposition
DMF	dimethylformamide
EDTA	ethylenediaminetetraacetic acid
ET	enzyme thermistor
FAD	flavin adenine dinucleotide
FET	field effect transistor
FMCA	ferrocene monocarboxylic acid
GOD	glucose oxidase
HDH	hydrogen dehydrogenase
HK	hexokinase
ISFET	ion sensitive field effect transistor
LDH	lactate dehydrogenase
MES	2-[N-morpholino]ethane sulphonic acid
MIC	Minimal inhibitory concentration
MOS	metal oxide semiconductor
NEP	N-ethyl phenazinium
NHE	normal hydrogen electrode
NMA	N-methyl acridinium
NMP	N-methyl phenazinium
PCMH	p-cresolmethylhydroxylase
PMS	phenazine methosulphate
PTFE	polytetrafluoroethylene
PU	polyurethane
SCE	saturated calomel electrode
TCNQ	tetracyanoquinodimethane
TELISA	thermometric enzyme linked immunosorbent assay
TEP	thermal enzyme probes
THF	tetrahydrofuran
TMOS	thin metal film oxide semiconductor
TTF	tetrathiafulvalene

1

Unmediated amperometric enzyme electrodes

GEORGE S. WILSON and DANIEL R. THÉVENOT

1. Introduction

Since the development of the enzyme-based sensor for glucose first described by Clark (1) in 1962, there has been an impressive proliferation of applications involving a wide variety of substrates. These applications have recently been extensively reviewed elsewhere (2−4). These applications involve enzymes which catalyse redox reactions whose rates are made proportional to the analyte (substrate) concentration. Typically the progress of the reaction is monitored by measuring the rate of formation of a product or the disappearance of a reactant. If the product or reactant is electroactive, then its concentration may be monitored directly. The enzymes catalysing these reactions are typically oxidoreductases, but hydrolytic enzymes such as alkaline phosphatase can also be used if they produce an electroactive species. Because the species usually involved are small molecules, they can be monitored amperometrically without the need for a mediator, hence sensors based on these reactions are called 'unmediated amperometric enzyme electrodes'. The most common system by far involves the monitoring of the disappearance of oxygen or the appearance of hydrogen peroxide (2). Strictly speaking, a biosensor should be 'reagentless' meaning that no additional reagents need be added to make the sensor function. Oxygen is, of course, a reagent which is consumed in the reaction, but because it is usually already present in the sample, no reagent addition is required. A large number of enzymes use NAD^+ or NADH as a cofactor, which must be added to the solution. Regeneration of the cofactor within the sensor has not proven easy to implement and its electrochemistry is also not straightforward. See Chapter 3 for further details on NADH electrochemistry.

Although beyond the scope of this presentation, it is also possible to incorporate several enzymes into the same sensor. There are three reasons for doing this.

(a) Conversion of an analyte by a sequence of reactions into a form that can be conveniently detected electrochemically.
(b) Conversion of interferences in the sample into electrochemically or enzymatically inactive forms.
(c) Recycling of reactants to enhance enzymatic turnover.

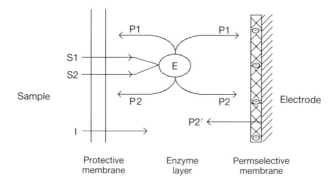

Figure 1. Schematic profile of amperometric enzyme electrode. S1,S2, substrates; P1,P2, products; P2′, product of electrochemical reaction of P2; I, interference (neutral or charged); E, enzyme.

The reader is referred to a recent review by Scheller (5) on this subject.

A schematic diagram of an enzyme electrode is shown in *Figure 1*. The electrode sensing element is usually constructed of platinum, but gold and various forms of carbon have also been used. Immediately adjacent to the electrode is the enzyme layer which is formed by the entrapment of the enzyme within a gel, by covalent glutaraldehyde-mediated cross-linking with a protein such a bovine serum albumin (BSA), or by covalent attachment to a support membrane. Direct attachment or adsorption of the enzyme on the electrode is also possible. The literature concerning the immobilization of biocatalysts is vast. An excellent recent monograph (6) should be consulted for details. Unfortunately, it is not possible to generalize about immobilization methods. Some methods work well with certain enzymes, but not with others. The methods which we present below are ones which are easily implemented, and which usually yield satisfactory results. The objective is to produce a layer of enzyme which is as thin as possible but with enzyme immobilized at the highest possible specific activity. Failure to do this results in sensors with poor sensitivity and long response times. The final component in the sensor is the outer (protective) membrane. This membrane serves several important functions. First, it is a protective barrier which prevents large molecules such as proteins from entering the enzyme layer. Biological fluids often contain catalase which could destroy the hydrogen peroxide produced in the enzyme layer thus leading to an erroneously low response. The membrane barrier will also prevent the leakage of enzyme into the sample solution. A properly chosen membrane exhibits permselective properties which are additionally beneficial to sensor function. At the applied potential corresponding to the oxidation of hydrogen peroxide it is also possible to oxidize a variety of amino acids as well as urate and ascorbate. However, if a membrane possessing a negative charge is employed, it can largely exclude anionic electroactive interferences (I^-) of this latter type. This is illustrated in *Figure 1* as a permselective inner membrane. If only one membrane is employed, the outer membrane can also be permselective. Finally the membrane can serve as a diffusional barrier for the substrate itself. Most enzymes follow some form of Michaelis—Menten kinetics which leads to enzymatic reaction rates largely non-linear with concentration. Enzyme-based sensors, however, are capable of linear dynamic ranges of several orders of magnitude because the response is controlled by diffusion through the membrane and not by enzyme kinetics. If the enzyme layer activity is low, then a thick membrane will be required to achieve good linear response. This will also lead to slow response. On the contrary, if the enzyme layer activity is high, a thin outer membrane is sufficient and a rapid response may be obtained. It is important to understand the basic principles of sensor function in order to optimize response characteristics, and Chapter 9 describes some of the theoretical tools for analysing biosensor performance.

2. Basic techniques

2.1 Enzyme immobilization

The proper functioning of an enzyme-based sensor is, of course, heavily dependent on the properties of the enzyme itself. There are a number of commerical sources of enzymes including: Boehringer-Mannheim, Calbiochem, and Sigma. Commercial sources of less common enzymes may be found by consulting Linscott's Directory of Immunological and Biological Reagents. For common enzymes such as glucose oxidase (GOx EC 1.1.3.4) several grades are available. In this case not only should the specific activity be considered but also the presence of impurities such as catalase. Oxidoreductases are very sensitive to immobilization and usually yield specific activities which are 5−20% of the soluble enzyme. Three types of immobilization techniques are illustrated below which will work for glucose oxidase and probably a range of other enzymes as well.

2.1.1 Entrapment behind membrane

This example will be illustrated for the preparation of a glucose sensor and is a modification of a previously published procedure (7). The sensor probe is shown in *Figure 2a* and a method for immobilizing due enzyme is given below.

Protocol 1. Physical adsorption of enzyme

1. Prepare a mixture of 24 g of cyclohexanone, 24 g of acetone and 1 g of cellulose acetate (39.8% acetyl content, available from Aldrich Chemical Co.).
2. Stir the mixture at room temperature until the cellulose acetate has dissolved and then cast a thin film on to the surface of the sensor probe. Allow the solvent to evaporate to leave a thin film on the surface.
3. Dissolve glucose oxidase (Sigma Type II, sp. act. 25 U mg^{-1}) in 0.1 M phosphate buffer, pH 7.4 to a final concentration of 25 mg ml^{-1}. Place 20 μl of this enzyme solution on top of the cellulose acetate membrane and allow the water to evaporate (5−10 min).

Unmediated amperometric enzyme electrodes

Protocol 1 *continued*

4. Cover the dried enzyme layer with a 1 cm square membrane of either collagen (Centre Technique du Cuir), polycarbonate (Nucleopore) or general purpose dialysis tubing (mol wt. cut off 12 000 – 14 000. Viscase Corporation) and fit it in place with an 'O' ring. The membrane should be held as tightly as possible without tearing it.
5. Trim off the excess membrane and place the probe in a 0.1 M phosphate buffer solution, pH 7.4 for 2 h before use.

Note: A suitable polycarbonate membrane has a 0.05 μm pore size and a 10 μm thickness.

A glucose sensor prepared in this way will be usable for several months if stored at room temperature in phosphate buffer.

2.1.2 Reticulation with glutaraldehyde

In some situations a higher loading of active enzyme can be obtained if the adsorbed enzyme is cross-linked with glutaraldehyde. Thus a minor variation of *Protocol 1* can be made.

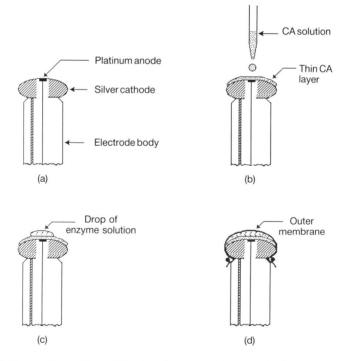

Figure 2. Preparation and immobilization of enzyme. (a) Sensor body; (b) application of CA membrane; (c) application of enzyme; (d) application of outer membrane.

Protocol 2. Reticulation of enzyme using glutaraldehyde

1. Prepare the probe exactly as described in *Protocol 1* up to step 3.
2. After the enzyme solution has dried add 10 µl of a 1% solution of glutaraldehyde. (A suitable source is Sigma Type I supplied as a 25% solution which should be stored frozen and diluted in water immediately before use.)
3. Allow the glutaraldehyde solution to evaporate and then fit the outer membrane as described in *Protocol 1*.

A variation on this reticulation procedure involves the mutual cross-linking of the enzyme with another protein such as BSA, Fraction V powder (Sigma). This procedure can lead to higher enzyme activity and greater stability.

Protocol 3. Reticulation of enzyme using glutaraldehyde and bovine serum albumin

1. Prepare the probe exactly as described in *Protocol 1* up to step 2.
2. Make the glucose oxidase solution up exactly as described in *Protocol 1* and also prepare a BSA solution (50 mg ml^{-1}) in the same phosphate buffer.
3. Mix 10 µl of each of the protein solutions and place the resulting 20 µl on the cellulose acetate membrane.
4. After 1−2 mins add 10 µl of 2.5% glutaraldehyde solution. The liquid layer should harden rapidly.
5. After a further 1−2 h fit the outer membrane as described in *Protocol 1*.

2.1.3 Covalent attachment to membrane

Covalent attachment procedures are more complicated but are especially useful in cases where the sensor is so small that the appropriate membranes must be fabricated directly on the sensing element. Under such conditions covalent procedues afford greater control over enzyme immobilization and give more stable and reproducible enzyme activity.

Collagen (acyl azide)

Collagen membranes (100 µm thick when in dry state, 300−500 µm when wet) can be obtained from the Centre Technique du Cuir, Lyon, France. Three to four membranes of 25 cm^2 total surface area can be derivatized simultaneously. This procedure has been described in detail by Thévenot and co-workers (8) and is presented below.

Protocol 4. Covalent attachment of enzyme to collagen membranes

1. Take three or four collagen membranes with a total surface area of ~25 cm^2.
2. Incubate the membranes in 50 ml of 100% methanol containing 0.2 M

Protocol 4 *continued*

   hydrochloric acid for 3 days at room temperature to convert the carboxyl groups to their methyl esters.
3. Wash the membranes carefully with distilled water.
4. Incubate the membranes in 100 ml of 1% hydrazine for 12 h at room temperature and then wash with water at 0°C.
5. Prepare a solution of 50 ml of 0.5 M potassium nitrite containing 0.3 M hydrochloric acid on an ice bath.
6. Immerse the hydrazine treated membranes in the nitrous acid solution for 15 min, then wash them with a 50 mM glycine-sodium hydroxide buffer pH 9.1.
7. Place the activated membranes in 5 ml of enzyme solution (1.5 mg ml^{-1}, at least 50 Units of activity) and store at 4°C for 12 h.
8. Wash the membranes with 0.1 M phosphate buffer pH 7.4.

Note: If the collagen is exposed for too long to the acid/methanol solution the membranes become extremely fragile.

Cellulose acetate (BSA – parabenzoquinone)

In this procedure the cellulose acetate surface is oxidized to produce aldehyde groups. These are then reacted with amine functionalities on the BSA to form a Schiff base linkage which is stabilized by borohydride reduction. This creates a BSA coating on the surface to which activated enzyme is coupled through amine functions using the *p*-benzoquinone coupling procedure. This procedure increases available functionalities on the CA surface and creates a BSA coating; for method, see *Protocol 5*.

Protocol 5. Covalent immobilization of glucose oxidase to cellulose acetate membranes

1. Dissolve 1.8 mg of cellulose acetate in a mixture of 20 ml of acetone and 3 ml of water.
2. Cast 1 ml of this solution on to a clean dry glass plate using a spreader (Touzart) and allow the solvent to evaporate for 1 min at room temperature.
3. Remove the membrane by immersing the glass plate in distilled water and floating it off. The resulting membrane is cut in to smaller pieces and stored at room temperature in water.
4. Suspend four membranes (each 2.5 cm square) in 100 ml of 0.1 M sodium periodate for 20 min at room temperature.
5. Wash the membranes in distilled water for 5 min then immerse them in 10 ml of a 10 mg/ml solution of BSA in 0.1 M borate buffer, pH 9 for 2 h.
6. Remove 9 ml of the BSA solution and add 4 mg of sodium cyanoborohydride (Aldrich). Incubate at room temperature for 2 h.

Protocol 5 *continued*

7. Wash the membranes in distilled water for 5 min and then store in phosphate-buffered saline at room temperature.
8. Recrystallize *p*-benzoquinone (Merck) from petroleum ether and prepare a solution of 15 mg/ml in ethanol.
9. Add 100 µl of the freshly prepared *p*-benzoquinone to 0.5 ml of a 20 mg/ml solution of glucose oxidase in 0.1 M phosphate buffer pH 7.4 in a tube covered by aluminium foil.
10. Incubate the mixture for 30 min at 37°C and then remove the excess *p*-benzoquinone by gel filtration through a Sephadex G-25 column (1 × 10 cm) equilibrated with 0.15 M sodium chloride and operating at a flow rate of 20 ml h^{-1}. Collect the pink-brown band that elutes in the void volume (2−3 ml). For further details consult ref. 9.
11. Suspend the BSA-cellulose acetate membranes in 2−3 ml of the activated glucose oxidase solution after adjusting the pH of the latter to 8−9 with 0.25 ml of 1 M sodium carbonate. Incubate at room temperature for 38 h.
12. Remove the membranes, wash them by stirring in 0.15 M potassium chloride solution for 24 h and then store them in phosphate-buffered saline pH 7.4 containing 1.5 mM sodium azide.

Note: The spreader has four channelled surfaces which yield films of 5, 10, 15 and 30 µm thickness. A 15 µm thickness is chosen.

Activated polyamide

Coulet and co-workers (10,11) have demonstrated the utility of activated nylon membranes for enzyme immobilization. Originally designed for immunochemical applications, Biodyne immunoaffinity membranes (120 µm thick, 0.2 µm pore diameter, Pall, Glen Cove, NY 11542 USA) have been successfully used for glucose oxidase and lactate oxidase immobilization.

Protocol 6. Covalent immobilization of enzyme to Biodyne membranes

1. Cut four 8 mm disks from a 120 µm thick, 0.2 µm pore size Biodyne membrane (Pall).
2. Immerse the membranes in 1 ml of a 1.5 mg/ml solution of the enzyme in 0.1 M phosphate buffer pH 7.4 and stir for 2 h at 4°C.
3. Wash the membranes twice for 20 min each time in 1 M potassium chloride and store in 0.1 M phosphate buffer pH 7.4 at 4°C.

Note: If lactate oxidase is immobilized on polyamide membranes its storage stability can be improved by the addition of 0.1 M potassium chloride, 10 mM magnesium chloride and 10 µM FAD to the storage buffer.

2.2 Protective membranes

In general the outer protective membrane must be compatible with the medium into which it will be placed and at the same time must allow the passage of substrates and analytes. For sensors with essentially planar active surface areas greater than about 1 cm^2, pre-cast membranes can be used. These have the advantage that their properties are generally more uniform than membranes deposited directly on the sensor from solution. They are also commercially available. Their disadvantage is that if the geometry of the sensor is not planar, then it may be difficult to position the membrane so that it is in uniform contact with the sensor surface. Failure to do so can cause the response characteristics to change with time. By contrast, deposition of polymer layers from solution produces a more adherent layer which can also accommodate a miniature or spherical geometry. Generally if pre-cast membranes are less than 10−15 μm thick, they cannot be manipulated without tearing. Therefore if a thinner membrane is desired, direct deposition is again the method of choice.

In the area of pre-cast membranes, three types are commercially available and easy to use. Collagen, a hydroxylic natural protein material is processed and cast into membranes. These membranes are easy to derivatize (see *Protocol 4*) and handle. At room temperature they work well, but at physiological temperature (37°C) they soften to the point of being unstable. They are compatible with biological fluids and exclude proteins, however, no obvious permselectivity is observed. Other sources of this material are FMC, Inc., and Sigma.

Synthetic materials available from Nucleopore Corp. in pre-cast form include polycarbonate membranes 'drilled' with neutrons to produce holes of uniform and controlled size. The 0.05 μm pore size, 10 μm thick membrane is the preferred material and it is strong and easy to handle. It exhibits no permselectivity for small molecules. The Biodyne immunoaffinity membranes mentioned in *Protocol 6* are a proprietary activated polyamide which reacts with amine functions on the protein. These membranes do not appear to exhibit significant permselectivity for small molecules.

A widely employed pre-cast material is cellulose acetate available as dialysis membrane and in the form of hollow fibres (Amicon). The polymer possesses some negative charge derived from the presence of residual carboxyl groups. At physiological pH these are ionized. Consequently CA membranes not only exclude proteins but are capable of retarding the transport of anionic species such as ascorbate and urate, two major electrochemical interferents, particularly when hydrogen peroxide is monitored. The actual selectivity depends on membrane thickness and preparation procedure, but *Table 1* gives some data that shows the magnitude of interferences for a glucose oxidase-based sensor particularly as applied to blood serum measurements. It is possible to cast membranes as thin as 5 μm using *Protocol 5*. If a spreader is not available, it is possible to cast a film on a glass plate by drawing a circle on the plate with a wax pencil. Depending upon the area chosen, a known volume of the CA solution is pipetted on the plate so as to produce a film of reproducible thickness. After the solvent evaporates, the film can be removed as described

Table 1. Substances interfering with glucose sensor response.

Substance	Interfering level[a] (mg dl^{-1})	Serum level (mg dl^{-1})
Acetone	26000	0.3 – 2.0
Beta hydroxybutyric acid	14000	–
Sorbitol	14000	–
D-xylose	730	–
D(–) adrenaline	110	–
Ascorbic acid	280	0.4 – 1.5
L(+) cysteine.HCl	100	0.9
D(–) fructose	5400	<7.5
d-Galactose	300	<20.0
Glutathione	100	28 – 34
d-Mannose	170	–
Tyrosine	160	0.8 – 1.3
Uric acid	400	3 – 7
Acetaminophen	1.5	–
Acetylsalicylic acid	167	–
Catechol	0.3	–
Sodium oxalate	11000	–
Heparin sodium	1800 U ml^{-1}	–
Sodium azide	360	–
Thymol	75	–
Epinephrine	–	18 – 26 ng dl^{-1}
Norepinephrine	–	47 – 69 ng dl^{-1}

[a] Corresponds to the level of interferent which would give an error of 5 mg dl^{-1} in an apparent glucose response. Measured with a Yellow Springs Instruments Model 2300 sensor. Interference data courtesy of YSI, Inc.

in the protocol. It should be pointed out that the membrane pore size is very dependent upon the solvent composition including water content, the rate of evaporation, the humidity and the temperature of the deposition environment (12). To obtain a reproducible product, it is therefore necessary to control these parameters carefully.

Three polymeric materials lend themselves well to deposition directly from solution: cellulose acetate, Nafion and polyurethane. The former material can be deposited on a sensor surface by dip coating using the solution described in *Protocol 5*. Nafion, a perfluorosulphonic acid ionomer made by DuPont is available in a low equivalent weight form (eq. wt 1000) which is soluble in low molecular weight alcohols. It can also be obtained as a 5% (w/w) solution in alcohol from Aldrich. By virtue of the negative charge created by the presence of sulphonate groups, a membrane fabricated from this material is capable of concentrating cationic species and excluding anions. These membranes have been deposited on surfaces in thicknesses as small as 1000 Å (13) and have been studied extensively in a variety of electrochemical applications (14,15). Nafion films can be deposited by dip coating with the 5% ionomer solution. Nafion films tend to adsorb proteins and other cationic species readily, and these may interfere with sensor response. Consequently, they are most effectively employed when coated with an external polymer layer such as polyurethane or CA which is more inert in this respect.

There have been numerous reports in the literature (16) involving the use of polyurethane (PU) as a biocompatible material. These have included its use as a protective membrane on implantable glucose sensors (17–18). Unfortunately, commercially available PU is produced in widely varying weight-average molecular weights which possess different functional groups. Thus the transport properties of this material as a film will differ considerably from source to source. Linear segmented aliphatic polyether-based polyurethane (EG80A or SG85A) is available from Thermedics Inc. A protective coating can be applied to a sensor by the following procedure.

Protocol 7. Application of a polyurethane protective coating

1. Prepare a mixture of 98% tetrahydrofuran and 2% dimethylformamide (v/v) and dissolve polyurethane in it to a final concentration of 4% (w/v).
2. Dip the tip of the sensor in the polyurethane solution and then remove it and allow the solvent to evaporate at room temperature.
3. Store the sensor in 0.1 M phosphate buffer, pH 7.4 for 2 days at room temperature prior to use.

PU is useful as an outer protective coating. If the PU solution is applied to a sensor surface which already has a cellulose acetate film and/or enzyme layer on it, care must be taken to ensure that the base films are not disrupted by the PU application. This is best accomplished by making one quick dip of the sensor into the PU solution. PU exhibits some permselectivity to small molecules and retards glucose access to the enzyme layer. This lowers sensitivity but leads to a sensor with a wide linear dynamic range (18).

2.3 Cell and sensor geometry

As most enzymatic reactions used for enzyme electrodes are irreversible, these biosensors deplete the substrate at their surface. Thus the supply of substrate to the sensor surface will be affected by hydrodynamic conditions in its vicinity. It is important to control solution flow by stirring (probe sensor) or by circulation of the sample solution (flow through sensor). Alternatively, the sensor may be rotated in the test solution. In all cases, the enzymatic membrane or layer, possibly also covered by a protective membrane, must be maintained in close proximity with the platinum working electrode. This positioning can be maintained with a screw cap (Radiometer Tacussel Type GLUC-1) or by the spacer of a modified liquid chromatography electrochemical detector (Radiometer Tacussel Type DEL-1) shown in *Figure 3* (19). In the latter case, solution is circulated through the cell using a Gilson Minipuls II peristaltic pump at a flow rate ranging from 0.1 to 2 ml min^{-1}. Auxiliary and reference electrodes are included in the sensor for control of applied potential. They are generally situated on the same side of the membrane as the working electrode in order to avoid resistive potential drop across the membrane.

Microsensors may be fabricated in a needle configuration (18).

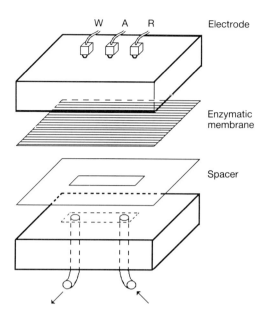

Figure 3. Schematic diagram of flow-through enzyme electrode. W, working electrode; A, auxiliary electrode; R, reference electrode.

Protocol 8. Fabrication of a needle-type microsensor

1. Heat the tip of a platinum wire (200 μm diameter, 10 cm length) to about 2450°C with an oxygen–butane microtorch to form a small sphere with a surface area of $1-2$ mm^2.
2. Seal the platinum wire into a polyethylene catheter (i.d. 0.3 mm, Biotrol Pharma) with epoxy cement.
3. Take a 23 gauge hypodermic needle (1.24 mm i.d.) and cut the end off square. Thread the catheter through the needle so that the platinum ball is held against its end and cover the inner half of the ball with epoxy cement.
4. Place the needle in a support and rotate it at 13 r.p.m. for 48 h to allow the epoxy to harden.
5. Wash the tip of the platinum sphere with trichloroethylene and then dip it in an ultrasonic bath of distilled water for 5 min to remove organic deposits.

Note: the area of the sphere can be conveniently determined with a micrometer.

The deposition of the enzyme and polymeric layers can then be carried out according to the procedures outlined in Sections 2.1 and 2.2.

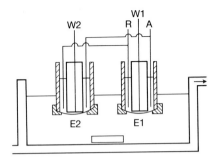

Figure 4. Schematic diagram of differential sensor apparatus. E1, enzymatic probe; E2, non-enzymatic (compensating) probe; W1,W2, working electrodes; A, auxiliary electrode; R, reference electrode.

2.4 Instrumentation

2.4.1 Principles

For the operation of a probe amperometric enzyme electrode, the following equipment is necessary: a thermostatted cell with magnetic stirrer, a sensor body, a potentiostat with amperometric readout, and a chart recorder. The temperature of the solution and of the sensor itself should be carefully controlled. The sensor sensitivity can vary by 3−10% per °C (20). Temperature also affects oxygen solubility and therefore the rate of enzymatic reactions that depend on oxygen such as those catalysed by oxidases. Control to ±0.1°C with a thermostatted cell and water bath is usually sufficient. It is important that all solutions i.e. buffer, standard and sample be brought to thermal equilibrium prior to measurement.

Protective outer membranes may not exhibit sufficient permselectivity to exclude all interfering electroactive species especially those which are uncharged. It is therefore necessary to make a differential measurement using dual sensors: an enzymatic and a non-enzymatic (compensating) element. Such a configuration has been found very useful for glucose determinations in food and clinical samples (8). The differential mode is illustrated in *Figure 4*. Sensors E1 and E2, corresponding to the enzymatic and compensating elements, respectively, are dipped into the same test solution. A potentiostat is used to maintain a potential difference of 650 mV between working electrodes W1 and W2 and the reference electrode, R, typically Ag/AgCl. The current outputs from W1 and W2 are first subtracted and then differentiated. Four time-dependent signals are thus available: $I1$, $I2$, $(I1-I2)$, and $d(I1-I2)/dt$ versus time. If a four electrode system such as that used for ring-disk voltammetry (Radiometer-Tacussel, BIPAD) or a dual electrode elelctrochemical detector for liquid chromatography (Bioanalytical Systems, Inc., LC-4B) or differential current amplifier (Radiometer-Tacussel, DELTAPOL) is unavailable, then it will be necessary to use two single potentiostats. There is a tendency for the two systems to interact electrically when common auxiliary and reference electrodes are used, so it is necessary to verify that the W1 and W2 outputs vary independently by addition of glucose and hydrogen peroxide or ascorbate, respectively. If only one potentiostat and working electrode is available, then a background current, Ib is determined in the absence of substrate. Subsequently analogous determinations of $I1$, $(I1-Ib)$ and $d(I1-Ib)/dt$ can be made.

Indicating (working) electrodes are generally made in the form of a platinum disk, wire, or foil. This material has been found to be better than gold or carbon for hydrogen peroxide detection. When oxygen is monitored, platinum or gold electrodes may be used alternatively.

The most widely used reference electrode is silver/silver chloride (Ag/AgCl) which can be prepared as a disk, ring or wire. Chloridation of the silver surface is easily performed by anodic oxidation under constant current or potential. Constant current is preferred because a more uniform and reproducible electrode usually results. The oxidation is carried out in 0.1 N HCl for 30 min at a current density of 0.4 mA cm^{-2} (21). Further details on the preparation of this type of electrode can be found in *Protocol 7* of Chapter 3.

When small indicating electrodes are used and consequently currents below 0.1 μA are measured, the usual three-electrode potentiostat (working, auxiliary and reference electrode) may be replaced by the simpler two-electrode (working and reference) system. In the latter case the reference electrode acts also as the auxiliary electrode and must maintain a stable potential even when current is passing through it. Possible potential variation in the reference electrode may be minimized by making its area ideally 4–5 times larger than that of the indicating electrode.

2.4.2 Assembly of low cost systems

The sensor body may be easily prepared by modifiction of a conventional gas electrochemical sensor. The hydrophobic membrane of a Clark-type oxygen sensor can be replaced or covered with enzymatic and protective membranes. An ammonia or carbon dioxide sensor can be used if the pH detector is replaced by a platinum disk working electrode covered with an enzymatic membrane (Model 8002-2 ammonia electrode—ABB Kent). Sensor bodies specifically designed for enzyme electrodes may be obtained from the following firms: GLUC-1 sensor (Radiometer-Tacussel), three-electrode probe Model 110708 (available in limited quantities—Yellow Springs Instruments, Inc.).

Although it is easy to design and build a single or differential potentiostat and amperometric unit using operational amplifiers, one may alternatively purchase these items. Potentiostats designed for liquid chromatographic detectors are quite suitable for this purpose becaues they can measure the small currents (microamperes to nanoamperes) characteristic of microsensors. Typical items are available from: Radiometer-Tacussel (Model PRG-DEL or PRG-GLUC) and Bioanalytical Systems, Inc. (Model LC4B). General-purpose or specifically designed workstations (22) can facilitate data acquisition and processing especially when numerous measurements are made.

2.4.3 Commercially available systems

There are integrated systems available which incorporate the sensor, readout device, temperature control, stirring and data acquisition into the same unit. These instruments are designed around specific analytes, but the enzymatic membrane provided can be replaced with one prepared using the methods described in Sections 2.1.1 and 2.1.3 allowing alternative analytes to be measured. Some sources of commercial instruments are: Yellow Springs Instruments, Inc. (Model 2000—detects hydrogen peroxide); Radiometer-Tacussel (GLUCOPROCESSEUR—detects hydrogen peroxide with differential electrodes); SERES (ENZYMAT—detects oxygen); SGI (MICROZYM-L—detects ferrocyanide).

2.5 Characterization of sensor response

2.5.1 Evaluation of sensitivity, stability, linearity and response

Calibration of the sensor is made by adding standard solutions of the analyte and is carried out in either of two modes depending upon whether the steady-state or dynamic response is measured as described below.

Protocol 9. Sensor calibration procedure

1. Dip the sensor into a thermostatted cell (at 37°C) containing 25 ml of buffer at the pH and ionic strength for optimal enzyme activity.
2. Apply the appropriate potential to detect the species of interest (+650 mV versus a Ag/AgCl reference electrode for hydrogen peroxide) and wait for the background current to stabilize. This takes typically about 20 min.
3. Add aliquots (25–125 μl) of standard analyte solutions (concentrations of 0.01–1 M) to generate a series of concentration steps.
4. Measure either the plateau current attained (steady-state response) or the maximum rate of change of the current from the derivative of the current–time curve (dynamic response).

The steady-state response is defined by the plateau reached in monitoring $(I1-Ib)$ or $(I1-I2)$ as a function of time. The dynamic response is obtained as the maximum of the current derivative i.e. $d(I1-I2)/dt_{max}$ or $d(I1-Ib)/dt_{max}$. The latter response can be measured more rapidly and thus improves overall sample measurement throughput. Dynamic response is proportional to the increase in substrate concentration in the reaction vessel (8) and this principle is frequently exploited in automated systems.

Steady-state responses are calculated by comparing the steady-state current either to the background current (Ib) in the absence of substrate or to the steady-state current corresponding to the previous addition. Thus either $I-Ib$ versus concentration (C) or delta.I/delta.C versus C or log C curves are plotted. Sensor sensitivity is best evaluated by measurement of delta.I/delta.C for each value of C in the cell. It is generally possible to measure the steady-state and dynamic responses over a large range of analyte concentration and successive substrate determinations are possible every 1−3 min by washing the sensor or rinsing the cell. If washing is required, it will be necessary to wait several minutes for the current response to return to the background levels. To facilitate comparison of sensors with different geometries, the observed sensitivity should be divided by the working electrode area (A), i.e. $(I-Ib)/A$ or (delta.I/delta.C)/A.

The limit of detection can be determined by comparison of background signal fluctuations and signal response. A signal/noise ratio of 2 is usually chosen as the limit definition. For very dilute solutions, i.e. 10−100 nM, the precision for the determination of substrate depends on the noise level, which is somewhat less for the steady-state than for the dynamic response. Probe electrodes generate less noise than flow through sensors because of the pulsation in flow rate created by the peristaltic pump in the latter case.

The linear range of the calibration curve is determined by plotting delta.I/delta.C versus C or by comparing delta.I/delta.C values for successive substrate additions. This method is much more definitive than plotting the usual calibration curves, $I-Ib$ versus C. The linear range usually extends over two orders of magnitude, between approximately 10 μM and 1 mM. When large working electrode areas are used, it is possible to obtain sensors linear between 100 nM and 3 mM (8). Response times are determined for each substrate pulse into the cell and are measured to 90 or 95% of the steady-state response. For the dynamic response the maximum value of the first derivative is used to define the response time. It is important to ensure that solution mixing or the time constant of the measurement electronics does not define the overall response.

Stability of sensor response may vary considerably depending on the sensor geometry, preparation method, and enzyme used. Sensors have been reported usable for periods of more than one year (23). How the sensor is stored and how frequently it is used will have important influences on its useful lifetime.

2.5.2 Assessment of specificity and interferences

Selectivity depends first upon the enzyme chosen. Most enzymes, except alcohol or amino acid oxidases, are very specific. Thus sensor E1 (*Figure 4*) yields a high selectivity for substrate. For example, glucose oxidase is 5×10^4 times more active with glucose than with other sugars such as fructose, lactose, or sucrose.

The main interference, therefore, is derived from electroactive species which can diffuse to the sensor surface to be oxidized. This is particularly a problem when the relatively high potential (+0.65 V versus Ag/AgCl) required to oxidize hydrogen peroxide is applied. This has led some investigators to suggest using a much lower potential (−0.4 V versus /Ag/AgCl) to monitor the oxygen decrease as a measure of substrate concentration. This can be done, but determination of the background signal is much more difficult. By use of the compensating electrode, E2, interference from such species as ascorbate, urate and tyrosine can be eliminated. For example, the selectivity coefficient for glucose-dependent hydrogen peroxide over nonenzymatically generated hydrogen peroxide is between 4×10^{-3} and 1.3×10^{-2} depending upon experimental parameters (8).

Assessment of selectivity is determined by comparing sensitivities (within the linear range of the calibration curve) for substrate and interferents. The parameters are calculated as $(\mathrm{delta}.I/\mathrm{delta}.C_{subst})/(\mathrm{delta}.I/\mathrm{delta}.C_{interf})$.

3. Conclusions

There are a large number of possibilities for the application of unmediated amperometric enzyme electrodes. A perusal of the literature will indicate that the vast majority of applications have involved either glucose or lactate as substrates. This is partly because the enzyme electrode is probably the method of choice in these cases and because these analytes are of considerable biomedial interest. One can envisage a variety of other applications particularly where the analyte is in the concentration range of millimolar to micromolar, where the sample matrix is complicated and where it is not desirable or possible to make a separation prior to analysis.

Acknowledgements

We would like to thank Dilbir S. Bindra and Marie-Bernadette Barrau for numerous helpful discussions concerning this manuscript.

References

1. Clark, L. C., Jr. and Lyons, C. (1962). *Ann. N.Y. Acad. Sci.*, **102**, 29.
2. Mosbach, K. (ed.) (1988). *Methods in Enzymology*. Academic Press, New York, Vol. 137.
3. Guilbault, G. G. (1984). *Handbook of Immobilized Enzymes*. Marcel Dekker, New York.
4. Turner, A. P. F., Karube, I., and Wilson, G. S. (eds) (1987). *Biosensors: Fundamentals and Applications*. Oxford University Press, New York.
5. Scheller, F., Renneberg, R., and Schubert, F. (1988). In *Methods in Enzymology*. Mosbach, K. (ed.), Academic Press, New York, Vol. 137, p. 29.
6. Mosbach, K. (ed.) (1987). *Methods in Enzymology*. Academic Press, New York, Vol. 135.
7. Sittampalam, G. and Wilson, G. S. (1982). *J. Chem. Ed.*, **59**, 70.
8. Thévenot, D. R., Sternberg, R., Coulet, P. R., Laurent, J. and Gautheron, D. C. (1979). *Anal. Chem.*, **51**, 96.
9. Sternberg, R., Bindra, D. S., Wilson, G. S. and Thévenot, D. R. (1988). *Anal. Chem.*, **60**, 2781.
10. Assolant-Volant, C. H. and Coulet, P. R. (1986). *Anal. Lett.*, **19**, 875.
11. Bardeletti, G., Séchaud, F., and Coulet, P. R. (1986). *Anal. Chim. Acta*, **187**, 47.
12. Kesting, R. E. (1985). *Synthetic Polymeric Membranes: A Structural Perspective*. 2nd edn., Wiley, New York.

13. Kristensen, E. W., Kuhr, W., and Wightman, R. M. (1987). *Anal. Chem.*, **59**, 1752.
14. Nagy, G., Gerhardt, G. A., Oke, A. F., Rice, M. E., Adams, R. N., Moore, R. B., Szentirmay, M. N., and Martin, C. R. (1985). *J. Electroanal. Chem.*, **188**, 85.
15. Gerhardt, G. A., Oke, A. F., Moghaddam, B., and Adams, R. N. (1983). *Brain Res.*, **290**, 390.
16. Folkes, M. J. (ed.) (1985). *Processing, Structure and Properties of Block Copolymers.* Elsevier Applied Science Publications.
17. Shichiri, M., Kawamori, R., Hakui, N., Yamasaki, Y., and Abe, H. (1984). *Diabetes*, **33**, 1200.
18. Sternberg, R., Barrau, M.-B., Gangiotti, L., Thévenot, D. R., Bindra, D. S., Wilson, G. S., Velho, G., Froguel, P., and Reach, G. (1989). *Biosensors*, **4**, 27.
19. Tallagrand, T., Sternberg, R., Reach, G., and Thévenot, D. R. (1988). *Horm. Metab. Res.*, **20**, 13.
20. Coulet, P. R., Sternberg, R., and Thévenot, D. R. (1980). *Biochim. Biophys. Acta*, **612**, 317.
21. Ives, D. J. G. and Janz, G. J. (1961). *Reference Electrodes: Theory and Practice.* Academic Press, New York, p. 179.
22. Thévenot, D. R., Tallagrand, T., and Sternberg, R. (1987). In *Biosensors: Fundamentals and Applications.* Turner, A. P. F., Karube, I., and Wilson, G. S. (eds), Oxford University Press, p. 705.
23. Thévenot, D. R., Sternberg, R., and Coulet, P. R. (1982). *Diabetes Care*, **5**, 203.

2

Mediated amperometric enzyme electrodes

H. ALLEN O. HILL and GURDIAL S. SANGHERA

1. Aim

The aim of this chapter is to outline the techniques involved, and the problems associated with, the design of redox-mediated amperometric enzyme electrodes. The practical aspects of the use of ferrocene and its derivatives as mediators for a variety of enzymes are discussed in detail with practical examples. In the final part of this chapter, the incorporation of redox proteins as mediators for amperometric enzyme electrodes is described and compared with non-biological mediators such as ferrocenes.

1.1 Introduction

The pioneering work of Clark and Lyons (1) stimulated progress in the use of immobilized enzymes for analysis (2−4). Electrochemical methods based on enzyme electrodes frequently involve monitoring either the consumption of oxygen, using a Clark electrode (5), or the formation of hydrogen peroxide using a platinum electrode (4,6).

The latter requires the application of a potential at which species, such as ascorbic acid and uric acid, are also electroactive. The effect of such interferences renders this approach to the analysis of clinical samples difficult without tedious pre-treatment. Similarly, in applications using oxidases, for which dioxygen is the physiological electron acceptor, the problem of its fluctuation, overcome in commercial analysers by pre-dilution of serum samples into oxygenated buffers, may be circumvented by choosing an alternative electron transfer acceptor. Usually the mediator is a low molecular weight species which shuttles (*Figure 1*) electrons between the redox centre of the enzyme and the working electrode (gold, platinum or carbon). Those that have been used include hexacyanoferrate(III), with enzymes having as substrates, glucose, hypoxanthine and cholesterol. However, a disadvantage of such systems is the inability to immobilize the mediator at the electrode. Organic dyes, widely used in spectrophotometric measurements, have a number of disadvantages for electrochemical use including ready autoxidation, instability upon reduction and pH-dependent redox potentials. Ideally a mediator for use in an electrochemical device

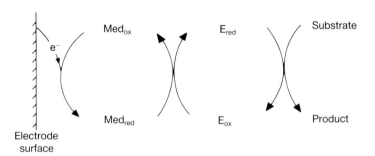

Figure 1. Reaction scheme depicting the role of a mediator for an enzyme-catalysed reaction where E is the enzyme and Med the mediator.

should react rapidly with the enzyme, exhibit reversible heterogeneous kinetics and possess a low overpotential for regeneration. Furthermore it should be stable with respect to pH, temperature, redox state and dioxygen.

The electrochemical technique of DC cyclic voltammetry provides an excellent method for probing enzyme–mediator interactions. The following section gives a broad view of cyclic voltammetry as applied to enzyme–mediator systems; a more detailed description of cyclic voltammetry may be found elsewhere (7).

2. Cyclic voltammetry

Conventionally, cyclic voltammetric experiments incorporate a three-electrode system avoiding current flow through the reference electrode. Instead the cell current is driven through a counter electrode, the solution and the working electrode (*Figure 2*). Placement of the reference electrode close to the working electrode minimizes potential drops due to high solution resistance; the design of three-electrode cells used throughout many of the examples discussed herein is described in the experimental procedure (see Section 3.1). For small electrodes, the current flowing in the cell will not be large and therefore will not perturb the reference equilibrium significantly; it may be convenient to use only a two-electrode configuration.

Consider the following reaction:

$$O + ne^- \rightleftharpoons R \tag{1}$$

where O = oxidized species; R = reduced species.

The potential applied, E, will control the concentration of the two redox forms in accordance with the Nernst equation:

$$E = E^\circ + (RT/nF)\ln[O]/[R] \tag{2}$$

providing the reactions are at equilibrium and the heterogeneous charge transfer kinetics are fast.

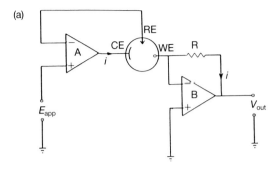

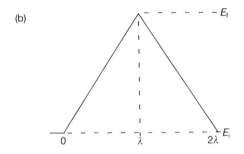

Figure 2. (a) A simple potentiostat circuit, using a control amplifier (A) and current follower (B) for a three-electrode system: WE, working electrode, CE, counter electrode and RE, reference electrode; (b) triangular wave form generated for cyclic voltammetry where λ is the switching time and E the applied potential.

In cyclic voltammetry, a triangular wave form describes the potential applied to a working electrode. The Faradaic current (i_f) will depend on the concentration gradient of O at the electrode surface (or Nernst diffusion layer) according to:

$$i_f = nF\,AD_o\,(d[O]/dx) \qquad (3)$$

where D_o is the diffusion coefficient of the electroactive species and A is the area of the electrode. As the potential of the electrode is made more negative, the surface concentration of species O progressively decreases, resulting in an increased concentration gradient and a larger current. As reduction occurs, the concentration of O at the electrode surface will be depleted and the current peak will decay. When the direction of the potential scan is reversed, a peak resulting from the re-oxidation of R, as governed by the Nernst relationship (equation 2) for a reversible system, is observed (*Figure 3a*).

Nicholson and Shain (7) numerically solved the relevant equations of charge transfer

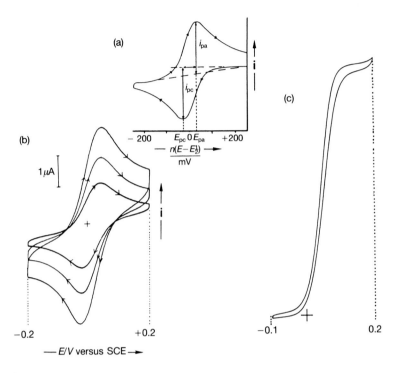

Figure 3. Cyclic voltammograms; (a) the relationship between anodic and cathodic (i_{pc} and i_{pa}) peak currents and the mid-point potential ($E_{1/2}$); (b) effect of increasing the potential sweep rate on the peak and capacitive current; (c) typical catalytic current voltammogram.

and mass transport for cyclic voltammetry and established the maximum cathodic current (i_{pc}) for a reversible system as:

$$i_{pc} = 0.4463 \; nF(nF/RT)^{1/2} \; D^{1/2} \; V^{1/2} \; C_oA \qquad (4)$$

where V is the potential scan rate and C_o the concentration of O in the bulk solution. Thus a reversible one electron transfer (*Figure 3a*) in cyclic voltammetry may be diagnosed by the following criteria:

(a) peak separation $\Delta E_p = E_{pa} - E_{pc} = 59$ mV ($59 \simeq 2.303 \; nF/RT$ at 298 K), where E_{pa} and E_{pc} are the anodic and cathodic peak potentials
(b) $I_{pa}/I_{pc} = 1$, where I_{pa} and I_{pc} are the anodic and cathodic peak currents;
(c) $I_{pa} \propto V^{1/2}$ (*Figure 3b*);
(d) E_{pa} is independent of V (*Figure 3b*).

2.1 Catalytic systems

A reversible electron transfer step, which is followed by a catalytic reaction, such as that involved in redox mediation with an enzyme, will obey the following scheme:

$$R + e^- \rightarrow O; \quad Z + O \xrightarrow{k_{cat}} R \quad (5)$$

where Z is a component which regenerates R, such as a reduced enzyme. The theoretical analysis for such a system has again been provided by Nicholson and Shain (7) for the case where [Z] ≫ [O]. From their treatment, different values for k_{cat}/a, where $a = nFV/RT$ and k_{cat} is the pseudo first order rate constant, are obtained. If k_{cat}/a, termed the kinetic parameter, is small then the cyclic voltammogram will approximate to that of a simple reversible electron transfer as depicted in *Figure 3a*. Conversely, if k_{cat}/a is large and the reduced species is continually replenished at the electrode then a limiting or plateau current is observed (i_L). Consequently no cathodic peak is observed since the concentration of species O in the Nernst layer will be negligible (*Figure 3c*). Numerically the limiting current (i_L) will be:

$$i_L = nFAC_o (D_o k_{cat})^{1/2}/1 + \exp[nF/RT(E-E_{1/2})] \quad (6)$$

In most practical examples, the limiting current is difficult to measure and an alternative procedure (7) is used to determine k_{cat}. The method involves the use of experimentally measured current ratios i_d/i_c, the ratio of the diffusion controlled current (i_d), to the catalytically enhanced current (i_c). From a theoretical working curve relating i_d/i_c to the dimensionless kinetic parameter $(k_{cat}/a)^{1/2}$, a range of k_{cat}/a values may be determined for several potential scan rates. A plot of k_{cat}/a versus V^{-1} gives a good estimate of k_{cat} and a plot of k_{cat} versus [Z] will be linear with a slope of k — the homogeneous second order rate of catalysis. A fully worked example of this kinetic treatment of voltammetric data is given in Section 3.2 for the ferrocene mediation of the reaction catalysed by glucose oxidase.

3. Ferrocene-based glucose sensor

Electrochemical investigations using DC cyclic voltammetry have shown (8–11) ferrocene, [$bis(\eta^5$-cyclopentadienyl) iron, $FeCp_2$] to be an excellent mediator for oxidase enzymes. Ferrocene itself exhibits a reversible redox couple with $E_{1/2}$ at +165 mV versus a saturated calomel electrode (SCE) and with the added advantage that the many derivatives available retain this redox centre with minimum change of its properties (*Figure 4*). The mediator replaces dioxygen as a cofactor for glucose oxidase (GOD) and many other enzymes and thus acts as an electron acceptor; the reduced form is then re-oxidized at a suitable electrode as in the following scheme:

$$\text{Glucose} + \text{GOD}_{ox} \rightarrow \text{GOD}_{red} + \text{gluconolactone} \quad (7)$$

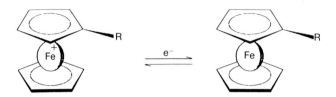

Figure 4. Structure of ferrocene, a one-electron mediator.

$$GOD_{red} + 2FeCp_2R^+ \rightarrow GOD_{ox} + 2FeCp_2R + 2H^+ \quad (8)$$

$$2FeCp_2R \rightleftharpoons 2FeCp_2R^+ + 2e^- \quad (9)$$

The design and response optimization of a ferrocene-based biosensor for glucose is described in the following section. However, the experiments outlined in this section may be extended to any given mediator/enzyme system.

3.1 Experimental methods

3.1.1 General materials

Electrolytes and substrates used must be of at least AnalaR grade and all solutions prepared using high purity water (Millipore). Enzymes generally may be used without purification directly from the manufacturer although some purification may result in a biosensor of increased stability. In this example, glucose oxidase, EC 1.1.3.4, type 2 from *Aspergillus niger* (supplied by Boehringer Mannheim), eluted as a single band on a fast protein liquid chromatography (FPLC). A mono-Q analytical column (Pharmacia) was used.

3.1.2 Electrochemical apparatus

DC cyclic voltammetry experiments can be performed with a two-chamber cell which has a working volume of 1 ml. *Figure 5* illustrates the type of cell that may be used with a 1 cm² platinum gauze counter electrode and an SCE as the reference electrode. For cyclic voltammetry experiments, a potentiostat (Oxford Electrodes) and a X-Y chart recorder (Bryans model 26000) are employed. For enzyme electrode calibration (in the amperometric mode) current/time curves are recorded on a Y-t recorder (Bryans model BS-271). The temperature of the cell is controlled with a thermocirculator (Churchill instruments).

3.1.3 Enzyme immobilization

Graphite electrodes are constructed by cutting a disk, 4 mm in diameter, and sealing it into a glass tube with epoxy resin with a connection to the external circuit by a wire, bonded with silver araldite, attached to the back of the electrode. After electrode pre-conditioning at 100°C for 40 h, and cooling in air, enzyme and mediator may be immobilized by the following steps.

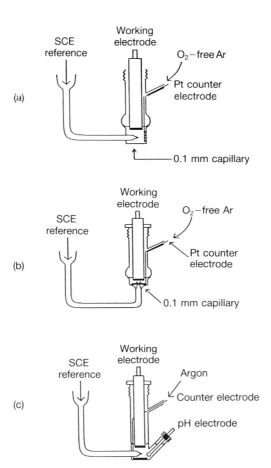

Figure 5. Three types of electrochemical cell for cyclic voltammetry: (a) 0.3–0.5 ml; (b) 0.2–0.3 ml; (c) with a combination pH electrode.

Protocol 1. Construction details for a glucose sensor

1. Deposit 15 μl of a 0.1 M ferrocene solution (in toluene) on to the electrode surface and allow it to air dry.
2. Place the electrode in 1 ml of a 0.15 M solution of 1-cyclo-hexyl-3-(2-morpholinoethyl) carbodiimide-p-methyltoluenesulphonate in 0.1 M acetate buffer pH 4.5 and incubate for 80 min at 20°C.
3. Thoroughly rinse the electrode in water and then place it in a stirred solution of 0.1 M acetate buffer pH 4.5 containing 12.5 mg ml^{-1} glucose oxidase. Incubate at 20°C for 90 min.

Protocol 1 *continued*

4. Rinse the electrode with the same buffer, cover it with 0.03 μm polycarbonate membrane (Nucleopore) and store in buffer containing 1 mM glucose.
5. Prior to use condition the electrode to give a stable current by maintaining its potential at +160 mV (versus SCE) in a solution of 7 mM glucose for 10 h.

3.2 Analysis and optimization of performance

DC cyclic voltammetry can be employed to determine the rate of electron transfer between different ferrocenes and the glucose oxidase/glucose system. The former exhibits reversible one-electron transfer voltammograms at a pyrolytic graphite electrode. *Figure 6a* shows a typical cyclic voltammogram for 0.5 mM ferrocene monocarboxylic acid (FMCA) in the presence of glucose. For this system the peak separation, ΔEp is about 60 mV at 298 K and $i_p/V^{1/2}$ is constant (i_p is the peak current and V the potental scan rate). From *Equation 4*, a plot of experimentally determined peak current values against the square root of potential scan rate, the diffusion coefficient of FMCA can be determined as about 3×10^{-6} cm^2 sec^{-1}. The change in the voltammogram upon the addition of glucose oxidase (10.9 μM) can be clearly seen (*Figure 6a* and *b*). The peaks completely disappear and instead a large catalytic current flows at oxidizing potentials; this current is due to the regeneration of ferrocene according to *Equations 7–9*.

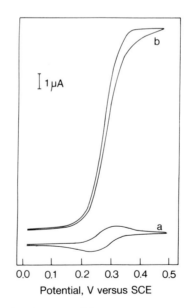

Figure 6. (a) DC cyclic voltammogram of ferrocene monocarboxylic acid (0.5 mM) in the presence of glucose, 50 mM, scan rate 1 mV sec^{-1}; (b) as for (a) with the addition of glucose oxidase, 10.9 μM. (Reproduced from ref. 8 with permission.)

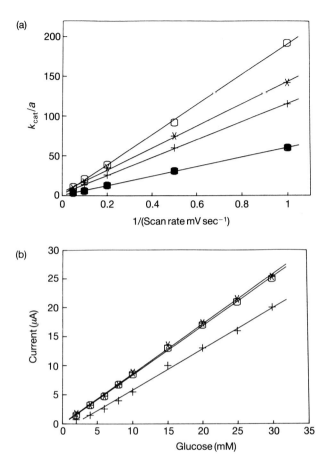

Figure 7. (a) The kinetic parameter, k_{cat}/a, as a function of the inverse scan rate (V^{-1}) for glucose oxidase concentrations of (■) 10.9, (+) 20.6, (✳) 29.3, (□)37.2 μM (Reproduced from ref. 8 with permission). (b) Calibration curve for the glucose electrode in (✳) argon-, (□) air- and (+) dioxygen-saturated buffer. (Reproduced from ref. 8, with permission.)

Quantitative kinetic information may be obtained by making use of the working curve relating i_c/i_d to the kinetic parameter k_{cat}/a (see Section 2.1). Thus i_d, the diffusion current, is the mean of $i_{pc}/V^{1/2}$ (constant for a reversible system) and i_c, the catalytic current, is calculated from the increase in current upon the addition of enzyme. From a set of i_c/i_d data points, values of k_{cat}/a may be determined for several potential scan rates. Thus the calculated values of k_{cat}/a are then re-plotted against the inverse of scan rate (V^{-1}), for a series of glucose oxidase concentrations as depicted in *Figure 7a*. From the slope of each curve, a good estimate of k_{cat} for each glucose oxidase concentration is obtained. A plot of k_{cat} as a function of glucose oxidase concentration will have a slope equal to the second order homo-

Table 1. Rates of the oxidation of glucose oxidase by the ferricinium derivative of the following ferrocenes (measured at pH 7.0 and 25°C).

Ferrocene derivative	E^o (mV)[a]	$10^{-5} k_s$ (M^{-1} sec^{-1})
1,1'-dimethyl	100	0.77
(Ferrocene)	181	0.26
Vinyl	250	0.30
Carboxy	289	2.01
1,1'-Dicarboxy	403	0.26
(Dimethylamino)methyl	386	5.26

[a] Relative to a saturated calomel electrode.

geneous rate constant (k) for the reaction between FMCA and glucose oxidase (in this example $k = 2.01 \times 10^5$ M^{-1} sec^{-1}). This simple kinetic analysis provides an excellent method for choosing a good mediator for a given enzyme system. A range of ferrocene derivatives and the rate of glucose oxidase oxidation determined as outlined above are given in *Table 1*.

For the design of a practical enzyme electrode, other criteria must also be considered (Section 1.1). Most importantly the solubility of the reduced form of the ferrocene derivative in aqueous solutions must be low to aid entrapment within the electrode. In this case 1,1'-dimethylferrocene proves to be the best mediator in terms of oxidation of the enzyme and desirable physical characteristics for immobilization. The commercial device described in Section 3.2.2 incorporates 1,1'-dimethyl-3-(1-hydroxy-2-aminoethyl)ferrocene.

3.2.1 Response optimization

1,1'-dimethylferrocene has a mid-point potential $E_m = +100$ mV versus SCE and consequently for amperometric experiments the electrode is poised at +160 mV. The enzyme electrode is calibrated over the range 1–30 mM glucose (stirred solutions). *Figure 7b* depicts a typical calibration curve with the background current (~1.5 µA) substracted and a linear response range of 0–30 mM glucose. Above this concentration the response is non-linear, becoming insensitive to additional amounts of glucose above 70 mM.

An important application of a glucose biosensor is in the clinical assay of whole blood and therefore the sensor is tested in air, argon and oxygen-saturated buffer (*Figure 7b*). Whilst there is very little difference between the response in air and under argon, under pure oxygen, there is a significant difference. However, since whole blood contains less than 200 µM dioxygen, this is not thought to be a problem.

Analysis of buffered solutions containing glucose (7 mM) and a range of metabolites commonly found in blood shows that only ascorbic acid at 0.13 mM gives any increase in current. Other parameters such as pH, temperature, stirring rate, etc., must be optimized prior to application of the sensor to real samples.

Figure 8. The ExacTech ferrocene-based glucose meter.

3.2.2 Glucose analysis of whole blood

Assay of plasma samples with the enzyme electrode are compared with results obtained with a Yellow Springs Instrument (YSI) glucose analyser routinely used in hospitals. The latter device also incorporates glucose oxidase but is based on the detection of hydrogen peroxide in pre-diluted plasma. Results for a sample size (n) of 23 gives a correlation coefficent between the two assays of 0.98 (12).

An assay of whole blood, after the addition of heparin as an anti-coagulating agent, when compared with plasma glucose levels measured using the enzyme electrode yields a correlation coefficient for the two assays of 0.99 (for $n = 10$).

The transformation of the ferrocene-based enzyme electrode for glucose from the laboratory bench to a hand-held commercial device has been achieved and the instrument marketed world wide as the ExacTech glucose meter (13). The meter comprises a pen-sized potentiostat (length 136 mm), weighing less than 30 g, with an LCD display for glucose reading (*Figure 8*). The electrode, a disposable test strip, incorporates an immobilized layer of glucose oxidase and 1,1'-dimethyl-3-(1-hydroxy-2-aminoethyl)ferrocene, coated with a hydrophilic membrane to attract the blood sample (one drop). Each strip also contains its own reference electrode.

The test strips are calibrated on the ExacTech system with fresh whole capillary blood and the meter is able to operate between 18 and 30°C and 20−80% humidity, producing a glucose reading in mg dl^{-1} in 30 sec. An independent study comparing the glucose pen with a YSI system gave excellent agreement between the two methods (12). Furthermore, ExacTech results obtained from a group of nurses and technicians, compared with those obtained from patients showed good agreement, illustrating the ease of use of the meter. The ExacTech blood glucose meter kit includes an automatic lancing device and glucose control solutions for testing the meter, in an easy to carry travel case.

The design and optimization described for the glucose biosensor may be extended to any given enzyme which interacts with a mediator. Two further examples of redox-mediated biosensors illustrate:

- the adaptability of the glucose biosensor as a probe for the assay of the enzyme creatine kinase;

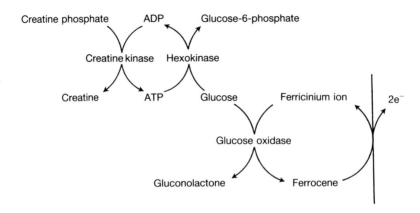

Figure 9. Reaction scheme depicting the creatine kinase assay *via* ferrocene-mediated glucose oxidase. For the ATP assay, CK was absent.

- the design of a cholesterol biosensor *via* three different ferrocene-mediated Systems.

4. Creatine kinase assay based on ferrocene

Creatine kinase (ATP: creatine *N*-phosphotransferase, EC 2.7.3.2) catalyses the reversible transfer of a phosphate residue from adenosine-5'-triphosphate (ATP) to creatine:

$$NH_2\overset{+}{C}(NH_2).N(CH_3)CH_2COOH + ATP \overset{CK}{\rightleftharpoons}$$
$$PO_3^{2-}\text{-}NH\overset{+}{C}(NH_2).N(CH_3)CH_2COOH + ADP \quad (10)$$

The product, creatine phosphate, represents an essential energy store for contraction, relaxation and transport of substances within muscle. Elevated levels of creatine kinase (CK) in blood are linked with acute myocardial infarction. Consequently the measurement of CK activity is of vital importance in post-myocardial infarction patients.

The glucose enzyme electrode incorporating ferrocene and described in Section 3.1.3 may be coupled to analytes other than glucose by using enzymes that compete with glucose oxidase for its substrate. Assays based on this strategy have been developed for ATP and CK according to the reaction scheme depicted in *Figure 9*. Thus the rate of current decrease at the electrode, will be proportional to the rate of creatine phosphate consumption from which CK activity may be estimated (14).

4.1 An ATP biosensor

An enzyme electrode, incorporating immobilized glucose oxidase and ferrocene, is used as described in Section 3.1.3. However, response optimization experiments

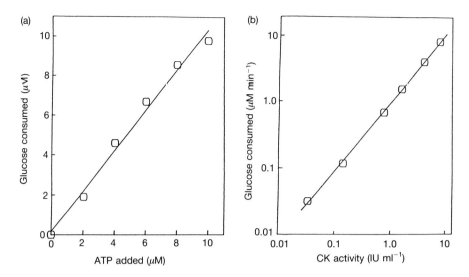

Figure 10. (a) Calibration curve for ATP (up to 10 μM) *versus* glucose consumption. (b) Calibration curve for creatine kinase *versus* glucose consumption.

are performed with all the enzymes and substrates outlined in *Figure 9* in a soluble form (in the absence of CK). Thus for optimization studies, the electrochemical cell (*Figure 5*) contains ferrocene monocarboxylic acid (FMCA), 0.5 mM in 25 mM Tris−HCl buffer, pH 7.0, 0.20 mM $MgCl_2$, 10 mM glucose and 10.9 μM glucose oxidase. Cyclic voltammetry of this reaction mixture yields a catalytic current (*Figure 6c*), and addition of 20 IU ml^{-1} of hexokinase (HK) produces no change in the shape of the voltammogram. However, addition of ATP, to a final concentration of 10 μM, results in the complete disappearance of a catalytic current. A voltammogram indicative of the reversible one-electron redox reaction of FMCA reappears (*Figure 6a*).

The addition of ATP results in the phosphorylation of glucose to glucose-6-phosphate (*Figure 9*) thus removing substrate from solution; hence no catalytic current is observed. The situation may be reversed simply by the addition of excess glucose.

The glucose enzyme electrode can be calibrated for ATP by subsequent additions of 2 μM ATP to the electrochemical cell in the presence of 20 IU/ml^{-1} HK and 10 mM glucose. Typically a linear response between 0−10 μM ATP (*Figure 10a*) is obtained and exhibits good agreement with the amount of glucose consumed.

4.2 Estimation of creatine kinase activity

Creatine kinase can be assayed by addition of CK (500 IU l^{-1}) to a solution containing glucose (20 mM), ADP (5 mM) and creatine phosphate (20 mM). From the initial rate of current decrease, produced as a result of adding CK, the rate of glucose consumption can be calculated. *Figure 10b* shows a graph of the rate of glucose consumption versus the CK activity in a buffered solution, for which a correlation coefficient of 0.99 was calculated. The assay of CK can be improved by co-immobilization of the HK with the glucose oxidase at the fabrication stage of the glucose electrode.

5. Ferrocene-mediated cholesterol biosensor

A good practical example of the versatility of ferrocene-mediated amperometric electrodes is provided by the variety of ways in which cholesterol may be determined. The following example illustrates three routes for cholesterol determination leading to the development of a cholesterol biosensor (15).

5.1 Dehydrogenase biosensor

Dehydrogenase enzymes require NADH as a cofactor which may be coupled to the enzyme diaphorase, known to exploit ferrocene as a mediator according to the scheme depicted in *Figure 11a*. The construction of the sensor is described below.

Protocol 2. Construction details for a dehydrogenase cholesterol sensor

1. Prepare a solution containing 0.1 M NAD^+, 0.2 mM hydroxymethyl-ferrocene, 5 Units of cholesterol esterase, 5 Units of cholesterol dehydrogenase in 1 ml of 50 mM Tris adjusted to pH 8.5 with phosphoric acid.
2. Add 0.1 ml of serum and incubate for 10 min at 20°C.
3. Measure a cyclic voltammogram to confirm the reversible one-electron oxidation of the ferrocene derivative.
4. Add 50 Units of diaphorase and remeasure the cyclic voltammogram. In this case a catalytic current is obtained and a kinetic analysis of the system using the methods described in Sections 2.1 and 3.2 yields a second order rate constant for the oxidation of diaphorase of 8.8×10^5 M^{-1} sec^{-1}.
5. When operated in the constant potential amperometric mode a linear dependence of the current upon the concentration of cholesterol (or its esters) is obtained up to concentrations of 1 mM.

Note: this electrode is very sensitive to inhibition by the surfactants that are used to liberate the cholesterol esters from lipoprotein complexes in the serum.

5.2 Oxidase biosensor

The use of a specific flavoprotein oxidase, coupled to a ferrocene derivative, seems the most obvious approach in the development of a cholesterol biosensor. However, an investigation of cholesterol oxidase (COD) showed that only COD purified from

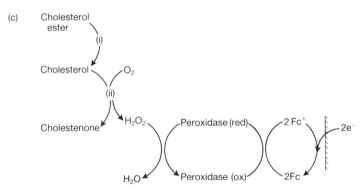

Figure 11. Three routes for the design of a ferrocene-mediated cholesterol biosensor. (a) Dehydrogenase assay; (b) oxidase assay; (c) peroxidase/oxidase assay. (For a more detailed explanation, see text.) (Reproduced from ref. 15, with permission.)

Schizophylum commune could be re-oxidized by ferrocene according to the scheme depicted in *Figure 11b*.

Assay of cholesterol using this enzyme employs the following procedure:

Protocol 3. Construction details for an oxidase cholesterol sensor

1. Prepare a solution containing 50 mM citrate buffer pH 5, 0.1 M potassium chloride, 0.2 mM hydroxymethylferrocene and 50 U cholesterol oxidase.

Protocol 3 *continued*

2. Add varying concentrations of cholesterol in propanol.
3. Poise the potential of the electrode at +400 mV (versus a saturated calomel electrode) and measure the current. A linear response up to cholesterol concentrations of 10 mM is obtained. Cyclic voltammetric investigations yield a rate constant for the reaction of the enzyme with hydroxymethylferrocene of 0.6×10^4 M^{-1} sec^{-1}.

The pH optimum of this enzyme is 4.5–5.5, rendering it unsuitable for the analysis of clinical samples of pH 7.4.

5.3 Peroxidase/oxidase biosensor

Cholesterol, oxidized by COD, generates hydrogen peroxide which may be detected by a peroxidase-based assay incorporating ferrocene (16), as depicted in *Figure 11c* and described below.

Protocol 4. Construction details for an oxidase/peroxidase cholesterol sensor

1. Mix thoroughly 3 g of carbon with 1.5 ml of 0.2 M pH 7.0 2-[N-morpholino]-ethane sulphonic acid (MES) in the presence of 5% cholate.
2. Add to the carbon paste 60 mg of hydroxymethylferrocene ensuring all the solid mediator is incorporated into the paste.
3. Finally dissolve in 1.0 ml of MES buffer cholesterol oxidase (2500 units); cholesterol esters (2500 units) and horseradish peroxidase 10 000 units and work into the paste uniformly.
4. Pack the paste into a well (3 mm) in a teflon sheath 7 mm in diameter and make electrical connection via a brass rod and conducting silver epoxy.

The second order rate constant for the reduction of HRP by the ferrocene is 1.33×10^4 M^{-1} sec^{-1}; in this assay the ferrocene acts as an electron donor to HRP rather than acceptor, as in the previously described systems.

5.3.1 Cholesterol biosensor construction

The previous section outlined three types of cholesterol assay each of which may be converted to a simple biosensor by immobilization of the enzymes and mediators involved. In this case the peroxidase-linked assay proved to be the most reliable and was used for the construction of a cholesterol biosensor.

The dehydrogenase-based assay exhibited extreme sensitivity to inhibition by surfactants, required to free cholesterol esters from lipoprotein complexes in whole sera. The alternative COD assay was limited by the very narrow pH range of the enzyme (pH 4.5–5.5); this is unsuitable for clinical samples (~pH 7.4).

Table 2. The principal classes of redox proteins.

Name	Characteristics
Cytochromes	Contain iron in a haem prosthetic group, e.g. cytochrome c involved in electron transfer in the mitochondrion.
Ferredoxins	Contain iron and sulphide in dimeric e.g. chloroplast [2Fe – 2S] ferredoxin, and tetrameric e.g. bacterial 2[2Fe – 2S] ferredoxins, involved in photosynthetic and nitrogen fixation electron transfer respectively.
'Blue' copper proteins	Contain copper bound to at least one cysteinyl residue in a distorted tetrahedral environment, e.g. plastocyanin and azurin involved in electron transfer in photosynthesis and perhaps in nitrite reduction.
Flavoproteins	Contain a conjugated organic moiety as the prosthetic group. They act as low-potential electron transfer proteins, e.g. flavodoxin.

The electrode exhibits linearity over the range 0–10 mM cholesterol. Results for cholesterol in serum compared against a standard clinical method yields a correlation coefficient of 0.993. The next step in the development of this cholesterol biosensor will be conversion to a commercial device.

6. Electrochemistry of redox proteins

Redox proteins comprise a range of biological mediators involved in a variety of processes from cell respiration to reactions in photosynthetic systems. The major types of redox proteins known are listed in *Table 2*; an important feature of these proteins is the nature and disposition of charged groups on the surface of the protein and the location of the redox centre with respect to the surface.

This subtle molecular architecture affords a great selectivity and specificity in the interaction of these proteins with each other and with enzymes. The structure (17) of cytochrome c (mol. wt 12 400) is illustrated in *Figure 12*. The Fe porphyrin (haem) centre is buried within the molecule; its solvent exposed surface corresponds to a very small proportion (0.06%) of the total molecular surface (18). The protein bears an overall positive charge of +9 due to an excess of basic lysine residues. There is also a significant dipole moment (~324 Debye) due to an imbalance in the spatial distribution of acidic side chains. A number of lysine residues are distributed around the solvent exposed haem edge and are thought to be the interaction domain of cytochrome c with redox partners.

6.1 Modified gold electrodes

For many years it was thought that reversible direct electron transfer between electrodes and proteins was not possible. Several reasons were presumed with the buried nature of the active site and irreversible adsorption of the protein resulting in denaturing on the electrode surface being mainly responsible. Modification of the surface of a gold electrode by adsorption of 4,4'-bipyridyl (19) provides a suitable

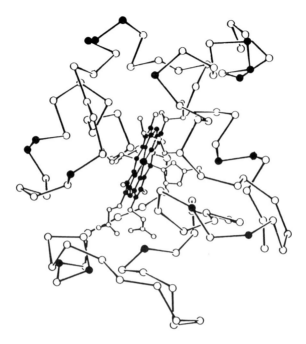

Figure 12. α-Carbon skeleton of horse heart cytochrome c. The haem group is shown here 'edge-on'.

surface for interaction with cytochrome c. It is important to note that the 4,4'-bipyridyl is not electroactive in the potential region under study and consequently does not act as a mediator.

The achievement of quasi-reversible electrochemistry of cytochrome c at 4,4'-bipyridyl modified gold electrode is a result of weak hydrogen bonding between lysine residues and pyridyl nitrogens at the modified electrode surface (20). Thus a transient protein−electrode complex, orientated to allow rapid electron transfer, results with the following sequence of events occurring (19);

(a) diffusion of cytochrome c to the electrode;
(b) binding of the protein to the surface, $k_{ads} \sim 3 \times 10^{-4}$ m sec^{-1};
(c) electron transfer, $k_{et} \sim 50$ sec^{-1}; and
(d) desorption of the protein, $k_{des} \sim 50$ sec^{-1}.

Following this work a comprehensive survey of 55 possible surface modifiers for protein electrochemistry at a gold electrode was carried out (21). The conclusions drawn from this work indicate the need for a bifunctional reagent $(X \sim Y)$, such that group X should bind to the electrode through N, P or S and group Y should be weakly

Table 3. A range of peptide modifiers for a gold electrode which act as *promoters* for electron transfer to cytochrome c and cytochrome b_5.

Protein	Peptide modifier i.e. promoter	k_s (cm sec^{-1})
Cytochrome c	LysCysOH	~5 × 10^{-4}
Mol. wt 10 500,	CysLysOH	1 × 10^{-3}
net charge +9,	GlyCysOH	~1 × 10^{-2}
haem edge	CysGlyOH	1 × 10^{-3}
exposed ~0.06%	CysGluOH	1 × 10^{-3}
Cytochrome b_5	ArgCysOMe	6 × 10^{-3}
Mol. wt 10 000,	GlyCysOMe	8 × 10^{-3}
net charge -9,	LysCysOH	3 × 10^{-3}
haem edge	CysLysOH	~1 × 10^{-2}
exposed 23%		

anionic or weakly basic for interaction with the binding domain of the protein.

More recently the range of modifiers has been extended to the use of amino acids and small peptides as modifiers and promoters of protein electrochemistry (22). These are based on sulphur-containing amino acids, cystine and cysteine; a few examples are listed in *Table 3*.

6.2 Graphite electrodes

Protein electrochemistry may also be carried out at carbon electrodes. Pyrolytic graphite, formed by deposition of carbon from the vapour phase, comprises layers of carbon atoms arranged in a polyhexagonal fashion. These polyhexagonal rings are stacked along the 'c' axis, that is at right angles to the plane of deposition, to give a hexagonal (ABAB....) lattice. This highly ordered structure gives rise to two distinct surfaces depending upon which plane is cleaved (23). The direct electrochemistry of cytochrome c at the distinct surfaces of graphite is illustrated in *Figure 13*. The parallel or basal plane provides a hydrophobic surface with ideally satisfied carbon valences and thus poor cytochrome c electrochemistry apparently results, *Figure 13a*. Polishing this surface ruptures the C—C bonds on the electrode surface producing an increase in the C:O ratio as determined by Electron Spectroscopy for Chemical Analysis (ESCA) (*Figure 13b*). Alternatively, cutting the graphite across the aromatic rings ruptures the strong C—C bonds which are immediately oxidized giving rise to a further increase in the C:O ratio (*Figure 13c*). The direct electrochemistry of positively charged proteins may therefore be carried out on an 'edge plane' graphite electrode.

Direct electron transfer between negatively charged proteins, such as plastocyanin, and a graphite electrode (edge plane) may be achieved with cations such as Mg^{2+}, Ca^{2+} and $Cr(Am)_6^{3+}$ (where Am is an amino compound) as promoters (22). In this context the promoter is a redox inactive species in solution which enables fast interfacial direct electron transfer of the redox protein.

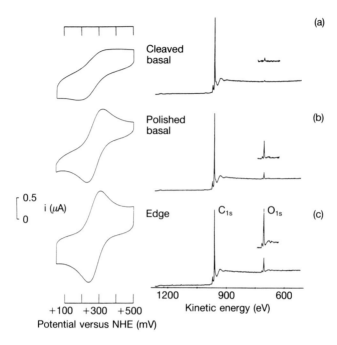

Figure 13. Cyclic voltammograms of 150 μM cytochrome c at basal and edge plane graphite. Also shown as the ESCA spectra (MgK$_{1,2}$ = 1253.6 eV).

The mechanism of cation-promoted electrochemistry may be understood if the edge plane electrode is visualized as having a high positive charge density as a result of binding the multivalent cations (*Figure 14*). Thus the interaction of the negative domain of a given protein and a graphite electrode results from binding of multivalent cations in 'cavities' generated within the transient interfacial structure. For a comprehensive review of protein electrochemistry the reader is referred to several reviews (24–6). The following section outlines the exploitation of redox proteins as mediators; in each example the redox protein is coupled to an enzyme. Thus the redox protein acts as a site-specific mediator for the enzyme under investigation.

6.3 Electrochemically driven respiration in mitochondria

There have been many investigations of the rates of electron transfer between isolated components of the respiratory chain, within mitochondria and in intact cells and organisms (27–9). Of the proteins involved or implicated, cytochrome c has been extensively studied: electron transfer to dioxygen *via* mitochondrial cytochrome oxidase occurs as shown in *Figure 15*. A *bis*(4-pyridyl)disulphide modified gold electron may be used to control the redox state of cytochrome c acting as a source or sink of electrons.

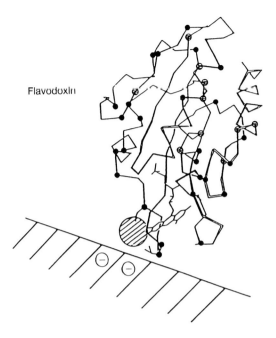

Figure 14. Schematic representation of the interaction of a negatively charged protein with the electrode *via* cation (•) promotion.

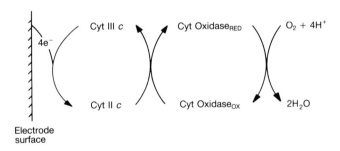

Figure 15. Reaction scheme depicting the catalysis by cytochrome oxidase of the reduction of dioxygen to water *via* cytochrome *c* mediation. (Reproduced from ref. 29, with permission.)

6.3.1 Practical considerations

The design of the bulk electrolysis cell is depicted in *Figure 16*. The working electrode, 1.9 cm² gold foil, is attached to the inside of the cell with epoxy cement. An additional removable disk-shaped gold foil electrode, area 0.39 cm², is sealed to the top compartment, with the rest of the cell holding a Clark-type oxygen

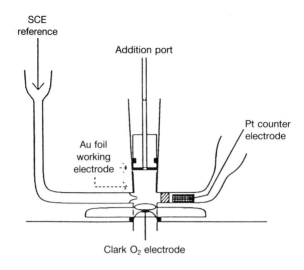

Figure 16. Bulk electrolysis cell for simultaneous measurement of current and dioxygen tension.

electrode. The chamber volume is ~500 μl and is stirred with a 7 mm PTFE-coated magnetic stirrer bar.

Prior to modification, the gold electrodes are polished with a 0.3 μm alumina water slurry on cotton wool. After thorough rinsing, the gold electrodes are immersed in a solution containing 2 mM *bis*(4-pyridyl)disulphide. The modified electrodes are then rinsed with reaction medium to give a surface which is long lived (>1 h). Experiments are conducted with a conventional three-electrode potentiostat at 20°C with a SCE reference and a platinum gauze counter electrode (as described in Section 3.1.2).

Horse heart cytochrome *c* (Type VI) obtained from Sigma is purified to remove unwanted deamidated and polymeric forms. The protein is purified by column chromatography on CM32 cation exchange resin (Whatman) according to the method of Brautigan *et al.* (30).

Rat liver mitochondria are isolated from adult rats according to the method of Weinbach (31). The reaction medium contains 25 mM sucrose, 0.1 M KCl, 10 mM $MgCl_2$, 1 mM EDTA, 10 mM KH_2PO_4 and 10 mM Tris at pH 7.2.

6.3.2 Results

According to the reaction depicted in *Figure 15*, as cytochrome *c* is reduced at a suitably modified electrode it is re-oxidized by cytochrome oxidase, in the presence of dioxygen. Cytochrome oxidase is not isolated from the mitochondria but may be coupled to cytochrome *c in situ*. Thus respiration by mitochondria is driven by cytochrome *c* electrochemistry with the protein acting as a site-specific redox mediator.

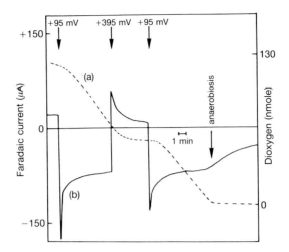

Figure 17. Dioxygen consumption (a), current uptake (b) *versus* time for horse heart cytochrome c and 2.0 mg ml^{-1} protein from rat liver mitochondria. Potentials *versus* NHE, as indicated. (Reproduced from ref. 28, with permission.)

Figure 17 illustrates the electrochemically driven respiration by rat liver mitochondria in the presence of exogenous cytochrome c. When the potential is stepped from +395 mV to +95 mV [versus normal hydrogen electrode (NHE)] a large reduction current is observed, with a corresponding increase in the rate of oxygen consumption which may be equated to the turnover of cytochrome c at the electrode, from which it may be calculated that four electrons are used per dioxygen molecule. On switching the potential back to +395 mV the current decays exponentially as the reduced cytochrome c is rapidly depleted from solution by both the modified gold anode and the cytochrome oxidase.

The electrochemical control of respiration in mitochondria illustrates the way in which protein mediators allow the investigation of the biological domains responsible for this process. Thus this type of *in vivo* biosensor enables the investigation of complex protein–enzyme interactions in biological systems.

6.4 A lactate biosensor

The following example directly compares the use of mediators coupled to the enzyme lactate dehydrogenase, LDH, (flavocytochrome b_2 from yeast). The direct electrochemistry of the redox protein cytochrome c may be coupled to LDH (32) according to the reaction depicted in *Figure 18*.

6.4.1 Experimental details

The electrochemical apparatus is as described in Section 3.1; the cell comprises a SCE reference, platinum gauze counter and a gold working electrode (see Section

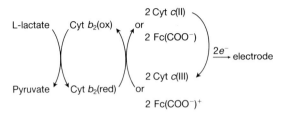

Figure 18. The oxidation of L-lactate to pyruvate by flavocytochrome b_2 via ferrocene or cytochrome c mediation. (Reproduced from ref. 32, with permission.)

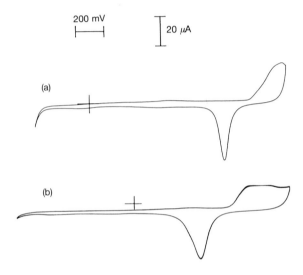

Figure 19. Steady state cyclic voltammograms of clean bare gold in (a) 1 M H_2SO_4; (b) 20 mM sodium phosphate, 100 mM sodium perchlorate, pH 7.0; both (a) and (b) at 20 mV sec^{-1} sweep rate.

3.1.2). A gold electrode polished with an alumina water slurry is further cleaned electrochemically in 1 M H_2SO_4, by cycling the potential between −900 mV and +1400 mV until a characteristic clean wave pattern for gold is obtained (*Figure 19*). Modification of the electrode is achieved by dipping the electrode in *bis*(4-pyridyl)disulphide for five minutes. After thorough rinsing the modified gold electrode exhibits quasi-reversible electron transfer in a solution containing 400 μM cytochrome c, (ΔE_p ~65 mV). The substrate for LDH, lactate, has no effect on the cyclic voltammogram when added to a final concentration of 10 mM. However, addition of 300 nM of the LDH produces a characteristic catalytic current (see *Figure 3c*). The enhanced anodic current is a result of the regeneration of the reduced

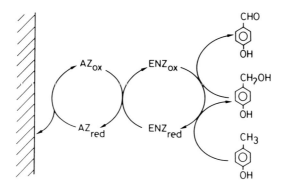

Figure 20. Reaction scheme depicting the electrosynthesis of p-hydroxy-benzaldehyde via azurin mediation. (Reproduced from ref. 33, with permission.)

cytochrome c(II) according to the reaction outlined in *Figure 18*. Kinetic data can be derived from the cytochrome c mediated to the enzyme LDH according to the procedure of Nicholson and Shain (7) (Sections 2.1 and 3.2). A second order rate constant k of 5×10^6 M^{-1} sec^{-1} for the mediation of cytochrome c to LDH is derived. In a parallel study, ferrocene carboxylate mediation of the enzyme, at a bare gold electrode yields a value of $k = 6.7 \times 10^6$ M^{-1} sec^{-1}.

The development of a lactate biosensor may proceed *via* ferrocene or cytochrome c mediation. Although in terms of cost and ease of mediator immobilization, the ferrocene approach, in the short term, is more feasible, the use of the more specific cytochrome c mediation is more desirable. A greater understanding of redox protein−enzyme interactions will enable the design of electrode surfaces such that the mediator may be dispensed with thus giving rise to a new generation of simpler direct electron transfer biosensors.

6.5 Electrosynthesis of *p*-hydroxybenzaldehyde

In the final example, the use of azurin, a blue copper protein, which is thought to be the natural electron acceptor for a number of hydroxylase enzymes from *Pseudomonas* bacteria is exploited to develop an electrosynthetic device (33). The enzyme *p*-cresolmethylhydroxylase (PCMH) is an effective catalyst for the oxidation of cresol and related phenols. The azurin-mediated electrochemical conversion of *p*-cresol into *p*-hydroxybenzaldehyde *via* PCMH occurs as shown in *Figure 20*.

Azurin (mol. wt 14 600) carries a net negative charge of 2−3 at pH 8.0, with the redox active copper site located near the surface of the protein where it is surrounded by hydrophobic residues. The crystal structure of azurin shows extensive charge pairing and there is no apparent sign of a binding domain for electron transfer.

The direct electrochemistry of azurin may be achieved at a gold disk working electrode modified with 2-pyridylmethylenehydrazinecarbothiomide; a 2 min dip in a 0.5 mM solution, followed by thorough washing, is employed. Pre-treatment of the gold electrode involves electrochemical cleaning as described in Section 6.4.1 until a clean wave pattern is obtained (*Figure 19*). Analysis of the dependence of the cyclic voltammetric response on scan rate and enzyme concentration according to the procedure of Nicholson and Shain (7) (as described in Sections 2.1 and 3.2) is used to derive an average second order rate constant between azurin and the enzyme of 9.0×10^4 M^{-1} sec^{-1} at 25°C pH 7.6.

The direct electrochemistry of azurin at the modified gold electrode can be used to drive the turnover of *p*-cresol methylhydroxylase with the near quantitative conversion of *p*-cresol to *p*-hydroxybenzaldehyde. Bulk electrolysis experiments are carried out with the cell described in Section 6.3 (*Figure 15*) with the oxygen electrode replaced by the base of the glass cell. The enzyme demonstrates high selectivity among the various isomers of cresol as no catalytic enhancement of current occurs with *o*- or *m*-cresol.

In the examples described in this section, redox proteins have effectively been acting as mediators of electron transfer to the relevant enzyme. The redox proteins exercise very good control over the electron transfer with regard to other components in the system. This feature is a quality lacking in small redox molecules which, although generally being much better electron transfer reagents, have no control of electron transfer either spatially or with respect to partners.

6.6 Horizons

It will be apparent that the ferrocenes act as very good mediators to a wide variety of enzymes. Of course there is a multitude of compounds which can act in this capacity so what, if anything, makes ferrocenes so special? Apart from qualitites such as fast electron transfer ratios (which are shared by other mediators), the principal quality of ferrocenes is that our organometallic colleagues have not been idle and there are literally *thousands* of ferrocenes. Most substituents have been introduced on to either or both rings and practically *all* ferrocenes retain their ability to act as mediators to some or even most redox enzymes. The potential of the ferrocene is, of course, dependent on the substituent but that can be put to good use by choosing one most suitable to the enzyme, substrate and/or conditions of interest. Even more important, the substituent can be chosen such that the compound formed is related to a suitable analyte which may be a drug or a component of a polynucleotide. An example (34) appeared where a ferrocene analogue of lidocaine was found to compete with the latter at an antibody site and hence a competitive assay was developed for the drug. Experimental details on this type of application may be found in Chapter 4. Other examples were readily apparent though many served to highlight one serious drawback of extending this simple method to other systems: it is relatively insensitive and when seeking to apply it to many drugs, or particularly components of DNA or RNA, methods of dramatically amplifying the signal without losing its inherent advantages, such as speed of response, must be found.

Other problems associated more with electrochemical methods in general, rather than a ferrocene-based one in particular, include improving the ratio of the observed to the background current. A way which shows considerable promise is that provided by the so-called microelectrodes of radius $50-0.1$ μm. One reason for their attractiveness is the increased signal-to-noise ratio observed but perhaps of more significance is simply their small size enabling at least one hundred microelectrodes to be formed on e.g. the surface of the working electrode of the ExacTech referred to above. Of course, not only would separate connections have to be made to each microelectrode (not too difficult, one would imagine, with the current skills available in microcircuity) but presumably different enzymes would need to be immobilized on each electrode surface. That is the challenge.

Acknowledgements

We thank all our colleagues for their considerable help and the Science and Engineering Research Council and MediSense Inc. for their support.

References

1. Clark, L. C. and Lyons, C. (1962). *Ann. N.Y. Acad. Sci.*, **102**, 29.
2. Fishman, M. M. (1980). *Anal. Chem.*, **52**, 185R.
3. Carr, P. W. and Bowers, L. D. (1980). In *Immobilized Enzymes in Analytical and Clinical Chemistry*. Elving, P. J. and Winefordner, J. D. (eds), John Wiley Inc., New York, Vol. 68, p. 254.
4. Guilbault, G. G. (1984). *Analytical Uses of Immobilised Enzymes*. Marcel Dekker, New York.
5. Updike, S. J. and Hicks, G. P. (1967). *Nature*, **214**, 986.
6. Guilbault, G. G. (1982). *Ion-Sel. Electrode Rev.*, **4**, 187.
7. Nicholson, R. S. and Shain, I. (1964). *Anal. Chem.*, **36**, 706.
8. Cass, A. E. G., Davis, G., Francis, G. D., Hill, H. A. O., Aston, W. J., Higgins, I. J., Plotkin, E. V., Scott, D. L., and Turner, A. P. F. (1984). *Anal. Chem.*, **56**, 667.
9. Cass, A. E. G., Davis, G., Green, M. J., and Hill, H. A. O. (1985). *J. Electroanal. Chem.*, **190**, 117.
10. Frew, J. E. and Hill, H. A. O. (1987). *Phil. Trans. R. Soc. Lond.*, **316**, 95.
11. Frew, J. E. and Hill, H. A. O. (1987). *Anal. Chem.*, **59**, 933.
12. Brown, E., Holman, R. R., Hughes, S., Matthews, D. R., Steemson, J., and Watson, W. (1987). *Lancet*, 778.
13. Manufactured by MediSense Inc., Abingdon, England and Cambridge, Mass., USA.
14. Davis, G., Green, M. J., and Hill, H. A. O. (1986). *Enzyme Microb. Technol.*, **8**, 349.
15. Ball, M. R., Frew, J. E., Green, M. J., and Hill, H. A. O. (1986). *Proc. Electrochem. Soc.*, **86–14**, 7.
16. Frew, J. E., Harmer, M. A., Hill, H. A. O., and Libor, S. I. (1986). *J. Electroanal. Chem.*, **201**, 1.
17. Dickerson, R.E., Takano, T., Eisenberg, D., Kallai, O. B., Samson, L., Cooper, A., and Margoliash, E. (1971). *J. Biol. Chem.*, **246**, 1511.
18. Stellwagen, E. (1978). *Nature*, **275**, 73.
19. Eddowes, M. J. and Hill, H. A. O. (1977). *J. Chem. Soc. Chem. Commun.*, 3154.
20. Albery, W. J., Eddowes, M. J., Hill, H. A. O., and Hillman, A. R. (1981). *J. Am. Chem. Soc.*, **101**, 7113.

21. Allen, P. M., Hill, H. A. O., and Walton, N. J. (1984). *J. Electroanal. Chem.*, **178**, 69.
22. Armstrong, F. A., Cox, P. A., Hill, H. A. O., Lowe, V. J., and Oliver, B. N. (1987). *J. Electroanal. Chem.*, **217**, 331.
23. Panzer, R. E. and Elving, P. J. (1975). *Electrochim. Acta*, **20**, 635.
24. Armstrong, F. A., Hill, H. A. O., and Walton, N. J. (1986). *Quart. Rev. Biophys.*, **18**, 261.
25. Frew, J. E. and Hill, H. A. O. (1988). *Eur. J. Biochem.*, **172**, 261.
26. Armstrong, F. A., Hill, H. A. O., and Walton, N. J. (1988). *Acc. Chem. Res.*, **21**, 407.
27. Cass, A. E. G. (1982). In *Inorganic Biochemistry*. Hill, H. A. O. (ed.), The Royal Society of Chemistry, London, Vol. 3, p. 183.
28. Coleman, J. O. D., Hill, H. A. O., Walton, N. J., and Whatley, F. R. (1983). *FEBS Lett.*, **154**, 319.
29. Coleman, J. O. D., Hill, H. A. O., Walton, N. J., and Whatley, F. R. (1984). In *Charge and Field Effects in Biosystems*. Allen, M. J. and Usherwood, P. N. R. (eds), Abacus Press, London, 1984, p. 61.
30. Brautigan, D. L., Ferguson-Miller, S., and Margoliash, E. (1978). In *Methods in Enzymology*. Colwick, S. P. and Kaplan, N. O. (eds), Academic Press, New York, Vol. 53, p. 128.
31. Weinbach, E. C. (1961). *Anal. Biochem.*, **2**, 335.
32. Cass, A. E. G., Davis, G., Hill, H. A. O., and Nancarrow, D. J. (1985). *Biochim. Biophys. Acta*, **828**, 51.
33. Hill, H. A. O., Hopper, D. J., Oliver, B. N., and Page, D. J. (1985). *J. Chem. Soc. Chem. Commun.*, 1469.
34. Di Gleria, K., Green, M. J., Hill, H. A. O., and McNeil, C. J. (1986). *Anal. Chem.*, **58**, 1203.

3

Conducting organic salt electrodes

P. N. BARTLETT

1. Introduction

Electrical conductivity is not a property frequently associated with organic compounds. Consequently electrically conducting organic materials originally attracted attention simply because they were unusual. The conducting charge transfer salts, with which this chapter is concerned, were first studied in the mid 1950s. The early work concentrated, not surprisingly, on the physics of the conduction process and included much work on the temperature dependence of the conductivity and the relationship between the crystal structures and the conductivity.

The electrochemistry of organic conducting salts dates from around 1979 when Jaeger and Bard began to look at organic conducting salts as electrodes and Kulys and his group first applied these materials as electrodes for bioelectrochemistry (1). Before we describe their applications in amperometric enzyme electrodes it may be helpful briefly to describe the crystal structures and the features necessary for reasonable electrical conductivity in these materials.

1.1 Conducting organic salts

In general conducting organic salts are made by the combination of a donor (D) and an acceptor (A). These species are typically planar molecules with delocalized π-electron density both above and below the molecular plane. Archetypal examples are the donor tetrathiafulvalene (TTF) and the acceptor tetracyanoquinodimethane (TCNQ). These and other examples are shown in *Figure 1*.

When one examines a number of different conducting salts several common features emerge. First, it is necessary to have a segregated stack structure. That is to say all the donor molecules should go into one type of stack and all the acceptor molecules into another. Systems in which the donors and acceptors are mixed up in the stacks are insulators rather than conductors. An example of this type of segregated stack structure is shown in *Figure 2*. Secondly, it is necessary that the donor (or acceptor) form a new aromatic sextet by the loss (or gain) of an electron, *Figure 3*. This ensures the mobility of the charge carriers within the stacks. In cases where both donor and acceptor form new aromatic sextets, as in TTF.TCNQ, then both donor and acceptor stacks contribute to the conductivity. Finally, it is important that there be partial charge transfer between the stacks. If charge transfer is complete, the charge carriers

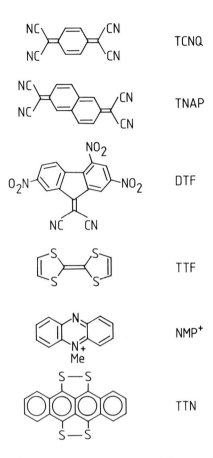

Figure 1. The structures of some common acceptor and donor molecules.

are again not mobile up and down the stacks and hence the material is an insulator. This requirement for partial charge transfer places some restriction on the choice of donor and acceptor pairs. It also follows from this that the conductivity is anisotropic. These materials have higher electrical conductivities parallel to the stacks than they do in the transverse direction. An excellent overview of organic conductivity can be found in the review by Perlstein (3).

From this discussion it will be clear that it is very difficult to predict *a priori* whether a given donor and acceptor will form a conducting organic salt or whether an insulator will result. For example, N-methyl phenazinium tetracyanoquinodimethanide (NMP.TCNQ) is a good electrical conductor ($\sigma = 143$ S cm^{-1}) but if the methyl group is exchanged for an ethyl group the corresponding compound, N-ethyl-phenazinium tetracyanoquinodimethanide (NEP.TCNQ), is an insulator ($\sigma = 10^{-9}$ S cm^{-1}). Fortunately quite a few conducting organic salts have been reported in

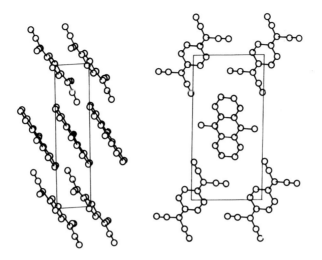

Figure 2. The crystal structure of NMP.TCNQ showing the separate stacks of NMP and TCNQ molecules. The NMP stack is disordered with respect to the orientation of the methyl group. Taken from ref. 2.

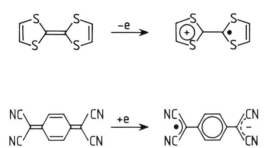

Figure 3. Structures of the oxidized and reduced forms of TTF and TCNQ respectively. In both cases a new aromatic sextet is formed. As a consequence both the TTF and the TCNQ stacks contribute to the conductivity in TTF.TCNQ crystals.

the literature and so it is possible to pick materials which we already know will be sufficiently conducting to be usable as electrodes.

2. Preparation of conducting organic salts

Of the conducting organic salts reported in the literature only a very few have been investigated as electrode materials or have been used to make enzyme electrodes. In general the compounds are readily prepared from the appropriate donors and

Table 1. Some conducting organic salts.

Salt[a]	Density (g ml^{-1})	Conductivity[b] (S cm^{-1})	Decomposition (°C)	Preparation method
NMP.TCNQ	1.44	2	192–193	Metathesis
NMA.TCNQ	1.33	2×10^{-5}	225–230	Metathesis
NMP.TCNQ$_2$	–	0.7	207–208	Recryst.
NMA.TCNQ$_g$	–	0.07	~273	Recryst.
NEP.TCNQ$_2$	–	0.83	220–225	Recryst.
TTF.TCNQ	1.62	500	–	Direct

[a] TTF, tetrathiafulvalene; TCNQ, tetracyanoquinodimethane; NMP, N-methyl phenazinium (PMS); NMA, N-methyl acridinium; NEP, N-ethyl phenazinium.
[b] All conductivities are for polycrystalline samples.

acceptors. We will concentrate here on the salts of TCNQ since these are the most commonly used.

The 1:1 conducting salts are prepared in good yields either by direct reaction of the donor and acceptor, for example

$$TTF + TCNQ \rightarrow TTF.TCNQ$$

or by metathesis between suitable donor and acceptor salts, for example

$$NMP^+CH_3SO_4^- + Li^+TCNQ^- \rightarrow NMP.TCNQ + LiCH_3SO_4$$

In addition to the simple 1:1 salts, in certain cases it is possible to get salts with the stoichiometry D.TCNQ$_2$. These can be prepared by recrystallization of the corresponding 1:1 salt from boiling spectroscopic grade acetonitrile in the presence of one equivalent of neutral TCNQ, for example

$$NMP.TCNQ + TCNQ \rightarrow NMP.TCNQ_2$$

Interestingly TTF does not appear to form a 1:2 complex of this type with TCNQ; thus TTF.TCNQ$_2$ is not known.

Table 1 lists some conducting organic salts along with their preparation. The two materials most frequently used to make enzyme electrodes are TTF.TCNQ and NMP.TCNQ and we will describe their preparation here. The preparation of other such materials follows very similar principles (4).

2.1 Preparation of TTF.TCNQ

TTF.TCNQ is readily prepared by direct reaction of the uncharged donor and acceptor in acetonitrile.

Protocol 1. Preparation of TTF.TCNQ

1. Weigh out separately 0.9 g of TTF (Fluka or Aldrich) and 0.9 g of TCNQ (Aldrich).

Protocol 1 *continued*

2. Dissolve the TTF and the TCNQ in separate 50 ml portions of hot spectroscopic grade acetonitrile (Fisons).
3. When both compounds have fully dissolved, mix the TTF and TCNQ solutions. A black precipitate should immediately form.
4. Cover the mixture and allow it to cool overnight with stirring.
5. The next day filter off the black crystals over vacuum.
6. Wash with a small amount of cold spectroscopic grade acetonitrile.
7. Wash with diethyl ether until the washings are colourless.
8. Dry under vacuum at room temperature to constant weight (overnight). This should give approximately 1.7 g (yield of ~90%) of a fine black crystalline product.

On mixing solutions of TTF and TCNQ, the salt precipitates from solution as a black microcrystalline powder. Both TTF and TCNQ are toxic and should be handled with care. Solutions of the two species are best freshly prepared as they can deteriorate on storage; in particular TTF solutions are light sensitive. In contrast, the conducting salt itself is very stable and can be kept for long periods (we have found no problems with material stored at room temperature for more than 2 years). It is important to use spectroscopic grade acetonitrile in the preparation. Other grades of acetonitrile contain impurities (such as acrylonitrile) which lead to adverse side reactions.

The TTF.TCNQ produced in this way is sufficiently pure to be used directly. The material may be recrystallized from spectroscopic grade acetonitrile. However, the components of the salt undergo some decomposition during this process and the resulting product is not significantly improved over the initial material.

2.2 Preparation of NMP.TCNQ

NMP.TCNQ is an example of a conducting organic salt which is most readily prepared by the metathesis route. The *N*-methylphenazinium ion is commercially available in the form of its methane sulphonate salt, $NMP^+.CH_3SO_4^-$. This material is frequently referred to as phenazine methosulphate, PMS. To prepare the conducting salt it is necessary to react the $NMP^+.CH_3SO_4^-$ with a suitable TCNQ salt. Li^+TCNQ^- is used for this purpose because it is soluble and easily prepared.

Protocol 2. Preparation of LiTCNQ

1. Weigh out 2.0 g of TCNQ (Aldrich) and 4.0 g of LiI (BDH general purpose reagent).
2. Reflux TCNQ and LiI in 200 ml of spectroscopic grade acetonitrile (Fisons) for 2 h.
3. Allow the resulting solution to cool overnight with stirring.

Protocol 2 continued

4. The next day filter off the purple solid over vacuum.
5. Wash the solid with cold spectroscopic grade acetonitrile until the washings are bright green.
6. Wash the solid with diethyl ether until the washings are colourless.
7. Dry at room temperature under vacuum to constant weight (overnight). This should give around 2.0 g of LiTCNQ ($\sim$100% yield).

Again note that TCNQ is toxic and must be handled with care. Spectroscopic grade acetonitrile should be used in the preparation as ordinary acetonitrile contains impurities which react with the TCNQ. Excess LiI is used in the reaction to remove the iodine produced as the soluble triiodide:

$$TCNQ + I^- \rightarrow TCNQ^{\cdot -} + 1/2\, I_2$$
$$I_2 + I^- \rightarrow I_3^-$$

LiTCNQ is quite soluble in a variety of solvents including water and lower alcohols. The purple solid dissolves to give a bright green solution when dilute or a blue solution at higher concentrations. This colour change is due to the formation of TCNQ$^-$ dimers in solution:

$$2\, TCNQ^{\cdot -} \leftrightarrow (TCNQ^{\cdot -})_2$$
$$\text{green} \qquad\qquad \text{blue}$$

The LiTCNQ produced by reaction of TCNQ and LiI can be used directly to prepare NMP.TCNQ.

Protocol 3. Preparation of NMP.TCNQ

1. Weigh out 1.03 g of LiTCNQ (prepared as described in the above protocol) and dissolve in 100 ml of refluxing absolute ethanol.
2. Weigh out 1.5 g of phenazine methosulphate (Aldrich) and dissolve it in 50 ml of hot absolute ethanol.
3. Add the PMS solution to the refluxing LiTCNQ solution. A black precipitate should immediately form.
4. Allow the solution to cool overnight with stirring.
5. The next day filter off the black crystalline product over vacuum.
6. Wash with a small quantity ($\sim$5 ml) of diethyl ether and leave to dry at room temperature overnight. This gives $\sim$1.7 g of crude product.

To recrystallize:

7. Dissolve the crude product in the minimum quantity ($\sim$200 ml) of hot spectroscopic grade acetonitrile (Fisons).
8. Filter the hot solution through a fluted filter paper into a hot flask.

Protocol 3 *continued*

9. Allow the filtrate to cool slowly overnight.
10. The next day filter off the purified product over vacuum as sparkling black crystals.
11. Wash with a small quantity of diethyl ether.
12. Dry at room temperature under vacuum to constant weight (overnight) to give ~1.5 g of purified material (75% overall yield).

For the metathesis reaction ethanol is used as the solvent. N-Methylphenazinium methane sulphonate is a mutagen and should be handled with care. All solutions should be freshly prepared. N-Methylphenazinium methane sulphonate is light sensitive and its solutions must be kept in the dark.

2.3 Preparation of single crystals

The preparative techniques described above are suitable for the preparation of relatively large quantities of conducting organic salts for use in the preparation of pressed pellet and other amorphous electrodes. The crystallites produced by these syntheses are generally very small, much too small to be used as single crystal electrodes or for single crystal conductivity measurements. For this reason other methods have been developed suitable for smaller scale preparation but yielding larger crystals.

There are a number of ways for growing large (typically $3 \times 0.2 \times 0.2$ mm) crystals of conducting organic salts. These all involve the slow crystallization of the salt. One approach is to allow solutions of TTF and TCNQ to slowly mix by diffusion. This can be achieved using a porous glass frit to separate the two solutions or by drawing out a tube to form a narrow constriction as shown in *Figure 4*. Solutions of TTF and TCNQ in spectroscopic grade acetonitrile are then carefully placed in the tube in two separate layers and allowed to mix by diffusion.

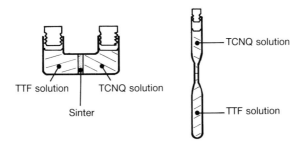

Figure 4. Two arrangements for the growth of large single crystals by slow inter-diffusion of solutions of TTF and TCNQ. It is important to stopper the solutions firmly to avoid evaporation of the solvent. With either arrangement the solutions should be kept undisturbed at a constant temperature in the dark.

Conducting organic salt electrodes

Protocol 4. Growth of single crystals of TTF.TCNQ

1. Prepare a number of tubes by sealing the ends of pieces of glass tube (0.5–1 cm diameter) and then drawing out the middle as shown in *Figure 4*.
2. Prepare saturated solutions of TTF (Aldrich) and TCNQ (Aldrich) in spectroscopic grade acetonitrile (Fisons).
3. Using a drawn out Pasteur pipette fill the tubes up to the middle of the constriction with the TTF solution. Take care not to get TTF on the walls of the top half of the tube.
4. Very carefully add the TCNQ solution on top of the TTF. Avoid any mixing of the two solutions or the entrapment of bubbles in the tube.
5. Stopper the tubes with rubber Subaseals and place them in a cool dark cupboard away from extraneous vibrations.
6. After about 6 months remove any large crystals by breaking the tubes. Filter the crystals off and wash with a little spectroscopic grade acetonitrile.

This is a slow process but after a period of 6 months or so, and with luck, some crystals of TTF.TCNQ may be recovered. The method is not totally reliable so it is best to make up a number of tubes, say a dozen, to increase the chances of success.

This slow inter-diffusion technique is unsuitable for NMP.TCNQ because the NMP^+ is not sufficiently stable in solution. In this case, slow recrystallization from spectroscopic grade acetonitrile can be used. It is important to ensure that the solution cools as slowly as possible to grow large crystals. One way to do this is to use a Dewar vessel. Using this technique, reasonably large crystals of NMP.TCNQ can be obtained. This approach also works for TTF.TCNQ but the crystals are not as large as those grown by diffusion. In particular although they are quite long (3–5 mm) they are also very thin. This is a common problem with the needle-like crystals of the conducting organic salts.

Finally these materials can also be prepared electrochemically. In this approach one of the components is generated at an electrode in the presence of the other. The two can then react together at the electrode to give the conducting organic salt. For example TTF.TCNQ can be prepared by oxidation of $TCNQ^{\cdot -}$ in the presence of TTF:

at the electrode

$$TCNQ^{\cdot -} \rightarrow TCNQ + e^-$$

followed by

$$TCNQ + TTF \rightarrow TTF.TCNQ$$

and the black crystalline product is formed at the electrode. This approach gives quite long crystals but again they can be very thin. Details for the method are given below.

Protocol 5. Electrochemical growth of TTF.TCNQ crystals

1. Clean two platinum flag electrodes (1 cm × 1 cm) by soaking overnight in a 1:1 mixture of Analar concentrated nitric and sulphuric acids—CARE. Rinse well with copious amounts of purified deionized water. Allow to dry.
2. Dissolve 0.030 g of LiTCNQ and 0.096 g of TTF (Aldrich) in a mixture of 40 ml of spectroscopic grade acetonitrile and 24 ml of Analar methanol (Fisons). Filter the resulting solution and place in a 100 ml three necked flask.
3. Place two 1 × 1 cm platinum flag electrodes in the solution and bubble with oxygen-free nitrogen for at least 20 min to remove any dissolved oxygen. Then seal the flask under nitrogen.
4. Pass a current of $10-50$ μA through the solution ($5-25$ μA cm^{-2}) using a voltage source and variable resistor. After about 2 h small crystals of TTF.TCNQ should begin to form on the anode (positive electrode). Leave at constant current for $1-2$ days.
5. When sufficient crystals have been grown carefully remove the anode and scrape the crystals on to a filter.
6. Wash the black crystals with spectroscopic grade acetonitrile and then diethyl ether.
7. Dry at room temperature under vacuum.

This approach can be used for other conducting salts. Further details of the electrochemical approach to the growth of single crystals of conducting organic salts can be found in the literature.

2.4 Other conducting organic salts

The preparation of other conducting organic salts follows the methods described above. Details of the preparations for a number of other materials can be found in the literature (4); some examples were given in *Table 1*. In many cases the donor or acceptor is not commercially available and must itself be synthesized before the salt can be prepared. Details of such syntheses are beyond the scope of the present chapter.

3. Electrochemistry

We now turn to the essential electrochemical instrumentation and apparatus necessary to use the conducting organic salts as electrodes in amperometric enzyme electrodes. This instrumentation is essentially standard electrochemical equipment that might be used for cyclic voltammetry or similar experiments (5,6).

3.1 Instrumentation

Amperometric electrodes can be designed to work using a two-electrode system or a three-electrode system (*Figure 5a*). The two-electrode approach is simple and

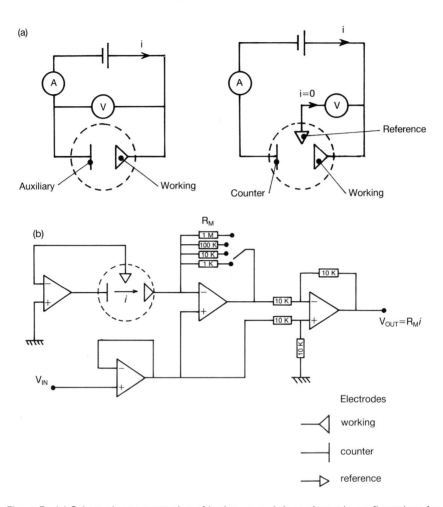

Figure 5. (a) Schematic representation of both two- and three-electrode configurations for amperometric measurements. In the two-electrode arrangement, the voltage is applied between the working electrode and the auxiliary electrode and the current flowing is measured. In the three-electrode set-up, an additional reference electrode is used. The potential between the working and reference electrodes is controlled. The resulting current flows between the working electrode and the counter electrode. (b) A simple design of potentiostat for three-electrode measurements. The choice of measuring resistor, R_M, determines the sensitivity of the current to voltage convertor.

requires little instrumentation. However it has a drawback. The potential that we apply between the two electrodes must drive the current through both electrode solution interfaces. Consequently we do not know how much of the potential drop is occurring at each interface; in other words we cannot be sure which electrode of the pair is primarily determining the current we measure or whether the current is determined by the balance between the two. This is not a very satisfactory situation for an amperometric device in which we want to use the current to measure the concentration of a particular species in solution. Having said this, it is possible to design the system to operate in a two-electrode mode if we are careful to ensure that the auxiliary electrode does not limit the current. We can do this by making sure there is a rapid electrochemical reaction which can occur at the auxiliary electrode. Having an auxiliary electrode of large area helps in this regard. The successful implementation of a two-electrode scheme requires some detailed knowledge of the electrochemistry of the system. Nevertheless it can work very well; indeed some successful commercial amperometric electrodes are based on two-electrode systems.

The drawbacks of the two-electrode system can be avoided if we use a third electrode; a reference electrode. We can then arrange to control the potential of our working electrode with respect to this reference electrode whilst measuring the current between the working electrode and the counter electrode (*Figure 5a*). Thus we have avoided the problem we had before because now we separate the two functions previously required of the auxiliary electrode in the two-electrode system. We do not have to worry about the potential drop at the second electrode because we are measuring potentials with respect to the reference electrode at which no current is flowing. From the point of view of the experimentalist in the laboratory, a three-electrode system is preferable for this reason.

To use a three-electrode system we must use a potentiostat to control the working electrode potential and allow us to measure the current. We can either use a commercial potentiostat or construct our own. *Figure 5b* gives the circuit for a simple potentiostat which is perfectly adequate for our purposes. In addition to the potentiostat we require a voltage source to provide the input potential to the potentiostat. Ideally we would also like a triangular wave generator so that we can sweep the potential when required.

If you do not wish to build your own potentiostat, a cheap commercial instrument is perfectly adequate for the low current measurements here and it is not necessary to spend thousands of pounds on this. All the results shown in this chapter were obtained using an Oxford Electrodes combined wave generator and potentiostat available for less than £1000. This unit contains a fixed voltage source as well as a triangular wave generator suitable for cyclic voltammetry measurements and is very well suited to the type of measurements described here. Another low priced potentiostat is the 1503 Thompson Microstat (Sycopel Scientific Ltd); this has a built in constant voltage source but does not include a triangular wave generator which must be purchased separately.

In addition to the potentiostat, some form of chart recorder is required to record the data. Ideally this should be an XY/t recorder so that both current−potential (cyclic voltammograms) (XY) and current−time responses at constant potential (Yt) can be recorded. The Gould Bryans 60000 series XY/t recorders are suitable and reliable but many other recorders can be used. Finally a digital voltmeter is very useful for monitoring the electrode potential and the current. *Figure 6* shows a general view of the apparatus used in the author's own laboratory.

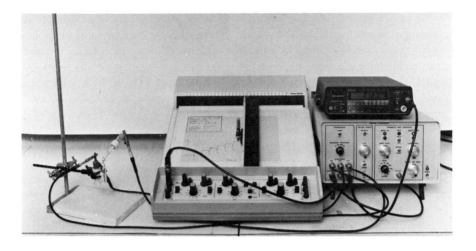

Figure 6. General view of the experimental apparatus used for enzyme electrode experiments, showing the electrochemical cell, Oxford Electrodes Combined Triangular Wave Generator and Potentiostat, Bryans chart recorder and Fluka digital voltmeter.

3.2 Electrodes

The fabrication of the conducting organic salt electrodes is described below. In addition to the working electrode we require a reference electrode and a counter electrode to complete the three-electrode system. The counter electrode is necessary to complete the circuit by passing an equal and opposite current to that flowing at the working electrode. To avoid problems in driving the current through the counter electrode it should ideally be large in area when compared with the working electrode. In addition we do not want the electrode to corrode and contaminate our solution. For these reasons the most common choice for the counter electrode is a large area (~ 1 cm^2) platinum gauze. This can be easily made by spot welding a suitable piece of gauze to a platinum wire as the contact. An alternative is to use a carbon rod as the counter electrode.

The reference electrode can be any electrode which has a stable, reproducible, redox potential. In practice the most commonly used electrodes are the saturated calomel electrode and the Ag/AgCl electrode. These are both commercially available, however they are also fairly easy to make in the laboratory. For our present purposes the reference electrode potentials need only be accurate to ± 1 mV which is readily attained. *Figure 7* shows the construction of a calomel electrode of the type used in our own laboratory. The most difficult part is the sinter. This must be sufficiently porous to allow good contact but not so porous that all the filling solution escapes. This can be achieved with a little practice. Alternatively a porous ceramic plug can be used. Details of the construction of the electrode are given below.

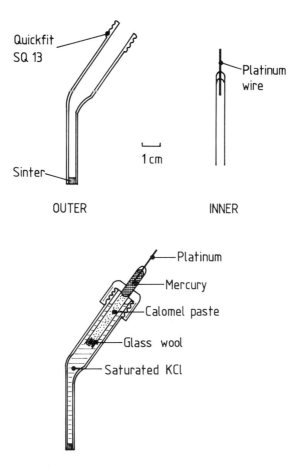

Figure 7. Construction of a saturated calomel electrode. The electrode consists of two parts; an inner and an outer containing a sinter. The assembled electrode is also shown.

Protocol 6. Construction of a calomel reference electrode

1. Thoroughly clean both the inner and outer parts of the electrode (*Figure 7*) in a solution of Decon 90 (BDH), prepared according to the manufacturer's instructions, overnight.

2. Rinse with copious quantities of purified deionized water to remove all traces of surfactant and then dry.

3. Solder a length of thick copper wire to the platinum contact protruding from the inner part, take care not to damage the platinum to glass seal.

4. Clamp the inner glass piece so that the platinum contact is at the bottom and place enough triply distilled mercury inside to cover the platinum wire.

Conducting organic salt electrodes

Protocol 6 continued

5. Prepare a calomel paste by grinding together calomel, Hg_2Cl_2, (Analar BDH), a small amount of triply distilled mercury, and some Analar KCl with a few drops of saturated KCl solution. The paste should have the consistency of toothpaste.
6. Pack the glass inner with the calomel paste, pushing it down so that there are no air bubbles and so that it makes good contact with the mercury inside the tube. The packing should come to within about 3–4 mm of the end of the tube.
7. Pack the end of the tube with a small amount of glass wool to prevent the calomel paste from falling out.
8. Fill the glass outer with saturated KCl and assemble the electrode, inserting the inner and screwing up the cap.
9. To make the electrode more robust, cut a short length of plastic tube to fit over the end of the protruding glass inner and then seal this to the copper wire using rapid Araldite (Ciba Geigy). This stops the platinum wire breaking off in use.
10. Place the completed electrode in saturated KCl and leave it to stabilize overnight. The potential should be stable and within ± 1 mV of a commercial calomel or another home-made electrode.
11. Do not allow the electrode to dry out. Store in saturated KCl and top up with saturated KCl as required. With care the electrode should last for 3 years or more.

It is important to take care over cleanliness at all stages.

The saturated calomel electrode is an excellent general purpose reference electrode. However it is not ideal if a very small reference electrode is required or if we wish to build the electrode into a self contained sensor (*vide infra*). In these circumstances the Ag/AgCl electrode is a much better choice since it can be prepared from a piece of silver, or silver plated platinum, wire.

Protocol 7. Construction of an Ag/AgCl reference electrode

Silver plating platinum wire

1. Clean the platinum wire thoroughly by soaking overnight in a 1:1 mixture of Analar concentrated nitric and sulphuric acids. Ensure that all traces of grease or organic matter are removed.
2. Rinse the wire with purified deionized water taking care not to touch it with your hands.
3. Prepare a silver plating solution of 0.25 g of $KAg(CN)_2$ (BDH) in 25 ml of purified deionized water. Remove excess cyanide by the dropwise addition of dilute silver nitrate until a faint cloudiness is produced.
4. Electrochemically deposit silver on the platinum wire using a current density

Protocol 7 *continued*

of 0.4 mA cm^{-2} for 6 h in a stirred solution. A snowy white deposit of silver should be produced.

5. Wash the electrode with purified deionized water.

Preparation of silver chloride

1. Prepare a solution of 0.1 M hydrochloric acid (Analar) in purified deionized water.
2. Clean the silver wire or silver plated platinum by dipping in 1 M Analar nitric acid solution for 10 sec.
3. Rinse the wire with purified deionized water and transfer it immediately to the 0.1 M hydrochloric acid solution.
4. Anodize the wire at a current density of 0.4 mA cm^{-2} for 30 min.
5. At the end of this time the wire should be evenly covered by a layer of silver chloride and should have a purple grey appearance.
6. Allow the electrode to stabilize overnight in 1 M KCl solution.
7. The next day, check the potential of the electrode against a commercial electrode or some other reference electrode by immersing both in saturated KCl.
8. Store the electrode in 1 M KCl[a] solution and do not allow it to dry out.

[a] AgCl will dissolve in any concentrated chloride solution:

$$AgCl\ (s) + Cl^- \rightleftharpoons AgCl_2^-\ (aq)$$

The silver plated platinum version is useful when glass is used as the insulating material since it is not possible to form good seals between silver and glass.

Table 2 gives data for some typical reference electrodes to allow comparison of their potentials.

Table 2. Reference electrodes and potential scales.

Electrode	Potential[a] versus NHE (V)	Potential[a] vs SCE (V)
Normal hydrogen electrode (NHE)	0	−0.2412
Saturated calomel (SCE) Hg/Hg$_2$Cl$_2$, KCl (saturated)	0.2412	0
Normal calomel (NCE) Hg/Hg$_2$Cl$_2$, KCl (1 M)	0.2801	0.0389
Sodium chloride saturated calomel (SSCE) Hg/Hg$_2$Cl$_2$, NaCl(saturated)	0.2360	−0.0052
Silver chloride Ag/AgCl, KCl(saturated)	0.197	−0.045

[a] All potentials refer to aqueous solution at 25°C.

3.3 Electrochemical cells and solutions

All electrochemical reactions are surface processes; electron transfer occurs at the interface between the electrode and the solution. As a consequence the processes can be very sensitive to impurities in solution particularly if they adsorb strongly to the electrode surface. It is therefore important to ensure that all glassware and solutions are as 'clean' as possible and as far as possible free from impurities. To this end Analar reagents and carefully purified water should be used to prepare all solutions as well as for rinsing electrodes and glassware.

Water may be purified either by double distillation from an all glass apparatus or by use of a commercial water purification system of the type used to deliver water for HPLC applications. One such system is the Millipore Milli-Q water purification system. This consists of a number of nuclear grade deionizing cartridges along with activated charcoal and microbiological filters. The water delivered from such a system is essentially free of dissolved organic compounds and ions.

All glassware should be thoroughly cleaned before use. This can be achieved by either soaking in a 1:1 mixture of concentrated Analar sulphuric and nitric acids followed by copious washing with purified water, or by soaking in Decon 90 (BDH), a commercial surfactant solution, followed by copious washing with purified water. It is important to take care to remove all traces of surfactant in the rinsing.

Where appropriate, working electrodes can be cleaned by polishing with a slurry, in purified deionized water, of alumina (Buehler) on a wad of absorbent cotton wool. The electrode is polished successively with 1 μm, then 0.3 μm diameter alumina. The electrode should then be rinsed with purified water before use. The reference electrode will generally not require cleaning but obviously should be rinsed with purified water before and after use. The counter electrode can be periodically cleaned by soaking overnight in a 1:1 mixture of Analar concentrated sulphuric and nitric acids.

In the preparation of buffer solutions for electrochemical measurements it is important to bear in mind a number of points. First, the reagents used should be as pure as possible to avoid problems of interference. Second, all solutions must be sufficiently conducting to pass the current. If this is not the case the electrochemical measurements will be distorted by the potential drop that is required to drive the current through the poorly conducting solution. This is generally avoided by using an inert background electrolyte such as KCl, KNO_3, K_2SO_4, NaCl, etc. at a concentration of 0.1 M or greater. This ensures a high solution conductivity. Thirdly, the buffer concentration must be sufficient to maintain the pH at the electrode surface where protons may be produced or used up in the reaction. For example, in an alcohol electrode based on alcohol dehydrogenase the reaction occurring is:

$$NAD^+ + CH_3CH_2OH \xrightarrow{ADH} NADH + CH_3CHO + H^+$$

One proton is produced for every ethanol molecule oxidized. It is therefore necessary to ensure that the buffer is able to mop up these protons as they are generated, otherwise the local pH at the electrode will not be held at the bulk pH value. In

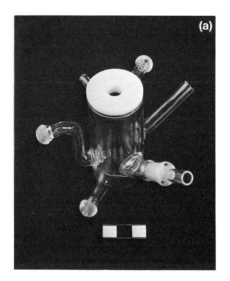

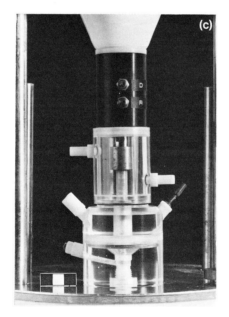

Figure 8. (a) A jacketed electrochemical cell with provision for bubbling with nitrogen to remove oxygen from the solution. The cell typically holds 15 ml of solution but is suitable for use with rotating disk electrodes. (b) A small volume (~3 ml) electrochemical cell for use with stationary electrodes. (c) A small volume (~3 ml) cell (Oxford Electrodes) for use with a rotating disk electrode. The cell is made from Perspex with the inside shaped to match the rotating disk flow pattern.

practice this means using a buffer near its pK_a, choosing one with fast proton exchange kinetics, and ensuring that it is used at a relatively high concentration, if possible 5−10 times that of the substrate. One last point to bear in mind about the buffer—it should not be electroactive in the potential range of interest!

We now turn to the electrochemical cell itself. This need not be particularly elaborate; a simple beaker fitted with a Teflon lid could be used. However, a number of factors should be borne in mind. Firstly it may be necessary to remove oxygen by bubbling an inert gas through the solution. Either argon or nitrogen can be used for this purpose. Note the gas should be pre-saturated with water by bubbling through a gas wash bottle containing the electrolyte solution otherwise there will be evaporation of water from the sample. Secondly, there should be some provision to thermostat the electrochemical cell. Finally, it is desirable to ensure that the products of the counter electrode reaction do not contaminate the solution. This can be achieved by placing the counter electrode behind a porosity 0 frit to prevent mixing. These various requirements are embodied in the cell design shown in *Figure 8a*. This is the type of general purpose cell used in our laboratory. It has an internal volume of approximately 15 ml and provision for passing oxygen free nitrogen either through or over the top of the solution during electrochemical measurements. The top of the cell is ground flat and fitted with a Teflon lid with a hole for the working electrode.

If smaller volumes of solution are required, because for example the species used are expensive or difficult to prepare, then a different design of cell is necessary. *Figure 8b* shows a small volume glass cell, capacity 2−3 ml, designed for use with a stationary electrode. When using rotating disk electrodes the solution hydrodynamics place an additional constraint on the design of small volume cells. *Figure 8c* shows a specially designed cell with a working volume of around 3 ml. The cell is made in Perspex and the inside is shaped to match the pattern of flow set up by the rotating disk. These cells and electrodes are commercially available (Oxford Electrodes).

4. Preparation of organic conducting salt electrodes

The conducting organic salts can be made into working electrodes in a number of ways. The choice of method depends upon the particular purpose. Thus the drop coating technique has the advantage of extreme simplicity but the disadvantage that the surface area is poorly defined. On the other hand packed cavity or pressed pellet electrodes are more robust and can be re-polished and re-used. Finally, for detailed mechanistic or electrochemical studies the single crystal electrode has the advantage of a defined reproducible surface. However it is fragile and not suited to sensor applications.

4.1 Drop coated electrodes

This is the simplest method for the preparation of an organic conducting salt electrode. It has the advantages of being simple, quick and not requiring very much material.

On the other hand it is in some ways the least satisfactory method because the electrode surface is poorly defined and very rough (it has the appearance of sandpaper).

The drop coating technique works best on glassy carbon electrodes. Platinum electrodes tend to give a rather patchy covering. The coating can be made more adherent if a small quantity of PVC is included in the solution. The drop coating solution should be freshly prepared since the conducting salts are frequently unstable in organic solution. Drop coating can be achieved by two different methods.

Protocol 8. Preparation of drop coated electrode

Method 1—No binder

1. Clean the electrode surface by polishing with a slurry of alumina in water.
2. Dissolve a small quantity of the organic salt in spectroscopic grade acetonitrile or other volatile organic solvent.
3. Carefully apply a drop of the solution to the surface of the electrode. With care the solution will wet the electrode surface (glassy carbon) but not spread onto the Teflon surround.
4. Place the electrode in a draught free atmosphere, for example under a large beaker, and allow the solvent to evaporate.
5. Visually inspect the electrode; it should be evenly covered with crystallites of the conducting salt.
6. If necessary repeat the process adding extra drops of solution until sufficient material is deposited on the electrode.
7. Wash the electrode with copious amounts of the appropriate buffer solution before use.

Method 2—PVC binder

1. Dissolve 0.015 g of PVC in 2 ml of purified tetrahydrofuran (THF). When the PVC is fully dissolved add the organic salt and allow it to dissolve.
2. Proceed as in Method 1 above from step (3).

Simple stationary disk electrodes can be easily fabricated from glassy carbon rod either press fitted into Teflon or sealed in heat-shrink Teflon (Korvex TOF available from Norton Performance Plastics). It is important to ensure a good seal around the outside of the carbon rod to avoid solution creeping between the carbon and the insulation. After assembly, and before application of the drop coat, the electrode should be polished to a flat mirror finish using successively finer grades of abrasive; 360 grit carborundum paper (Bramet) lubricated with water, 6 μm and 3 μm diamond lapping compound on a polishing pad (Engis), and then 1 and 0.3 μm alumina (Buehler) water slurry on cotton wool. Suitable stationary and rotating glassy carbon disk electrodes are also commercially available.

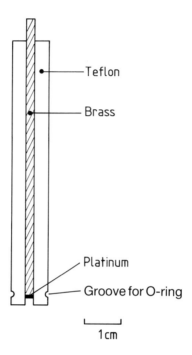

Figure 9. Construction of a cavity electrode. The dialysis tube membrane is held in place by a rubber 'O'-ring.

4.2 Packed cavity electrodes

The packed cavity electrode is a very convenient method for constructing a conducting organic salt electrode for use in an amperometric enzyme electrode. *Figure 9* shows a suitable design of cavity electrode. This is constructed by press fitting a platinum disk attached to a brass rod into a Teflon sheath. It is important to ensure that the seal between the Teflon and the platinum is watertight otherwise electrochemical corrosion of the underlying brass contact will occur. The exact dimensions of the cavity are not crucial but it is not advisable to make it too deep otherwise there may be problems with the conductivity of the packing. Typically a diameter of 5 mm and a depth of 1−2 mm works well. The groove around the edge of the electrode is used to hold a membrane in place using a rubber 'O'-ring.

Again it is important to prepare the packing mixture freshly before use as the components of the organic salt can deteriorate in contact with organic solvents. The best choice of solvent appears to be THF since this is a good solvent for the PVC binder. The THF should be Analar and should be purified immediately prior to use to remove the stabilizer, 2,6-di-tert-butyl-*p*-cresol, a hydroquinone derivative. This stabilizer is present to prevent the build up of explosive peroxides in the THF and can be simply removed by passing Analar THF (Fisons) down a column of activated alumina type 'O'.

It is important to clean the cavity before use in order to remove any traces of the old conducting salt or PVC:

Protocol 9. Preparation of packed cavity electrode

1. Dissolve 0.015 g of PVC in about 2 ml of purified THF.
2. Allow the THF to evaporate to leave a polymer solution with the consistency of runny honey. It should have a final volume of about 0.5 ml.
3. Add 0.09 g of the conducting salt in the form of small crystallites to the polymer solution and mix thoroughly. The resulting mixture should now have the consistency of putty.
4. Using a spatula smear the mixture into the cavity, taking care to pack it down well to ensure good electrical contact. Leave the surface standing proud of the surrounding Teflon insulation.
5. Allow the cavity to dry at room temperature for 2 h or longer and preferably overnight.
6. Trim the surface of the electrode back so that it is flush with the Teflon using a razor or fresh scalpel blade. It is possible to polish the electrode with alumina but this is difficult as the polymer bound material is quite soft.

The most common problem with the cavity electrode appears to be poor electrical contact between the platinum and the packing material. This problem can be overcome with a little practice. Once packed the electrode can be used for quite long periods (in excess of 2 months). However it is a good idea to repack the electrode periodically as the surface may become fouled. The cavity should certainly be repacked when changing enzyme systems to avoid contamination.

The cavity electrode can be re-used as required. To do this the cavity should be cleaned out and then soaked in THF to remove any traces of PVC. This should be followed by soaking in a 1:1 mixture of Analar concentrated nitric and sulphuric acids to clean the platinum. After rinsing in purified water and drying the electrode is ready to be repacked.

4.3 Paste electrode

Carbon paste electrodes, made from a mixture of carbon powder and an inert oil, are commonly used for electroanalytical applications. A similar approach can be used with the conducting organic salts. The paste can be stored and used with a cavity electrode of the type described above. The paste electrodes are a very good method for the preparation of organic conducting salt electrodes because they are easy to fabricate and give low background currents.

Protocol 10. Preparation of paste electrode

1. Thoroughly mix 6.5 mg of the conducting salt with 1 drop (~0.1 ml) of high temperature silicone oil (Aldrich) to give a stiff paste. This mixture can be stored for later use without observable deterioration.
2. Pack the electrode cavity with the paste.
3. Smooth off the electrode surface by wiping it across a sheet of white paper to give a flat working electrode surface.
4. Wash the electrode with the buffer solution to be used in the electrochemical experiments.

4.4 Pressed pellet electrodes

The pressed pellet electrode is another good method for making conducting organic salt electrodes. Although larger amounts of material are required the electrodes can be re-used with the surface of the pressed pellet being freshly polished each time. The pellets are prepared using a conventional infrared pellet press and dies of the type used to make KBr disks. The procedure is given below.

Protocol 11. Preparation of pressed pellet electrode

1. Select a 5 mm diameter die and ensure that it is clean and dry.
2. Place 0.06 g of a conducting salt in the die.
3. Evacuate the die on a water pump for 10 min.
4. Whilst still maintaining the vacuum apply a load of 2 tons to the die for 10 min.
5. Slowly release the pressure and carefully remove the pellet. Inspect for any flaws.

To turn the pressed pellet into a working electrode an electrical contact must be made to the pellet and the sides and back insulated. Electrical contact can be made with silver-loaded conducting paint (RS Components). Note that conducting silver epoxy resin (RS Components) is not recommended because the epoxy dissolves the pellet and gives a poor electrical contact. Alternatively a drop of mercury can be used to make excellent contact to the back of the pellet. *Figure 10* shows the construction of a pressed pellet electrode.

Once prepared, the pressed pellet electrode can be polished in the same way as a platinum electrode using a slurry of 0.3 μm alumina (Buehler) on a wad of absorbent cotton wool.

4.5 Single crystal electrodes

Although difficult to make and very fragile, single crystal conducting salt electrodes give much lower background current densities than polycrystalline electrodes and

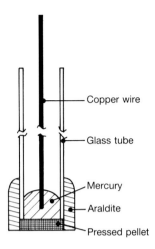

Figure 10. The construction of a pressed pellet electrode.

have well defined surfaces. For these reasons they are well suited to fundamental studies of the electrochemistry of the conducting salts. These electrodes have very small surface areas (around 0.1 mm^2) and are unsuited to analytical applications. Details on the preparation of the electrodes are given below.

Protocol 12. Preparation of single crystal electrode

1. Select a suitable single crystal.

2. Solder a 5 cm length of fine copper wire (~ 0.2 mm diameter) on to a 10 cm length of thicker copper wire (~ 1 mm diameter). A suitable source of the fine wire is a multicore cable.

3. Dip the end of the fine wire into conducting silver paint (RS Components) and carefully attach the crystal to the wire using the silver paint. Set aside to dry overnight.

4. The next day insert the wire into a glass melting point tube and pull it through until the crystal is about 0.5 cm from the end.

5. Wrap the other end of the wire around the tube and hold in it place. Then apply a small drop of rapid Araldite (Ciba Geigy) to the two ends of the tube to hold the wire in place. Take care not to allow the Araldite to come into contact with the crystal. Set aside to harden.

6. Under a microscope, insulate the remaining portion of fine wire, the silver contact, and the back and sides of the crystals with 3140RTV silicone rubber (Dow Corning) leaving one surface exposed. Set aside to cure.

7. Inspect the electrode under the microscope and check the electrode behaviour using cyclic voltammetry.

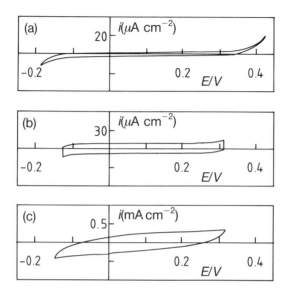

Figure 11. Typical cyclic voltammograms for three different types of TTF.TCNQ electrodes recorded at a potential sweep rate of 20 mV sec^{-1} in 0.1 M NaCl containing 0.15 M phosphate buffer at pH 7.4. (a) Single crystal, (b) paste electrode, (c) pressed pellet. Note that the charging currents are much larger at the pressed pellet electrode.

Note that Araldite will dissolve the crystals and so should not be used as insulation or, in the form of silver loaded Araldite, to make electrical contact to the crystal.

5. Electrochemistry of conducting organic salts

The most reliable method to check the behaviour of the conducting organic salt electrodes is to use cyclic voltammetry in buffer solution containing background electrolyte. Whilst the details are slightly different for each conducting organic salt the general principles remain the same in all cases (7). In this section we will concentrate on the behaviour of TTF.TCNQ and NMP.TCNQ.

First a brief word about cyclic voltammetry (6). In this technique we sweep the potential of the electrode between pre-set limits and record the resulting current as a function of the applied potential. The response that we observe depends on the sweep rate, our choice of potential limits, the composition of the solution and the previous history of the electrode. All of these factors are important for the conducting salt electrodes.

5.1 Electrochemistry of TTF.TCNQ

Figure 11 shows typical cyclic voltammograms for clean TTF.TCNQ pressed pellet, single crystal, and packed cavity electrodes recorded in pH 7 buffer solution. In

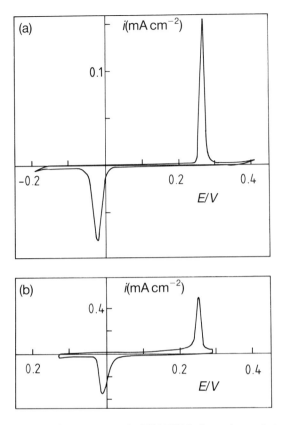

Figure 12. Typical cyclic voltammograms for TTF.TCNQ electrodes cycled outside the stable potential range. The sharp peaks are due to surface transformation of insoluble dissolution products. The voltammograms were recorded at a potential sweep rate of 20 mV sec^{-1} in 0.1 M NaCl, 0.15 M phosphate buffer at pH 7.4. (a) A single crystal electrode, (b) a paste electrode. Note that the coverages of surface material correspond to several thousand monolayers. In this case the surface peaks are due to TCNQ° and NaTCNQ.

this case we have been careful to stay within the stable range for the material. From the figure we can see that the cyclic voltammogram is essentially featureless over this range and under these conditions.

The stable potential range is set by the decomposition of the electrode by either oxidation or reduction of the electrode constituents. If we take the electrode to potentials more positive than about +0.4 V versus SCE the TCNQ$^{·-}$ is oxidized back to neutral TCNQ:

$$\text{TTF.TCNQ} \rightarrow \text{TTF}^+ + \text{TCNQ}^° + e^-$$

In aqueous solution the TCNQ° is insoluble and remains on the electrode surface.

Table 3. Surface redox processes on TTF.TCNQ[a].

Reaction	Peak potential (V versus SCE)
TCNQ → TCNQ$^-$(aq)	−0.04
TCNQ + K$^+$ → K.TCNQ	+0.04
K.TCNQ → TCNQ + K$^+$	+0.26
TCNQ + Na$^+$ → Na.TCNQ	−0.03
Na.TCNQ → Na$^+$ + TCNQ	+0.27
TTF → TTF$^+$(aq)	+0.20
TTF + Br$^-$ → TTF.Br	+0.08
TTF.Br → TTF + Br$^-$	−0.15
TTF + Cl$^-$ → TTF.Cl	+0.15
TTF.Cl → TTF + Cl$^-$	−0.20

[a] For further details see ref. 8.

The fate of the TTF$^+$ depends upon the solution composition. It can either dissolve into solution or form an insoluble deposit with the anions present in the solution. For electrolytes containing acetate TTF$^+$ is soluble, whilst for bromide, and to a lesser extent chloride, solutions insoluble TTF salts deposit on the electrode. These insoluble surface species are electroactive and show up in the cyclic voltammetry (*vide infra*).

If the clean electrode surface is taken to potentials more negative than about −0.2 V versus SCE the TTF.TCNQ becomes reduced giving insoluble TTF^o and TCNQ$^{·-}$:

$$TTF.TCNQ + e^- \rightarrow TTF^o + TCNQ^{·-}$$

This time the fate of the TCNQ$^{·-}$ depends upon the composition of the electrolyte. For lithium solutions the TCNQ$^{·-}$ is soluble but in the presence of sodium or potassium an insoluble TCNQ salt is deposited. Again these insoluble deposits are electroactive at the electrode surface.

We now turn to the electrochemistry of these surface species. *Figure 12* shows typical cyclic voltammograms for TTF.TCNQ electrodes which have been cycled outside the stable range in pH 7 phosphate buffer containing sodium chloride as the background electrolyte. There are two new sharp peaks in the cyclic voltammogram corresponding to surface transformations. The cathodic peak at −0.05 V versus SCE corresponds to the reaction

$$TCNQ^o(s) + Na^+(aq) + e^- \rightarrow Na.TCNQ(s)$$

The anodic peak at +0.27 V versus SCE corresponds to the reverse reaction

$$Na.TCNQ(s) \rightarrow TCNQ^o(s) + Na^+(aq) + e^-$$

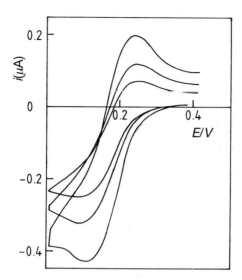

Figure 13. Cyclic voltammograms for the reduction of ferricyanide (2.23 mM) at a single crystal TTF.TCNQ electrode (area 0.092 mm^2) in phosphate buffer at pH 7.2. Sweet rates: 100, 50, and 20 mV sec^{-1}.

The surface peaks for other species are similarly sharp but occur at different potentials (*Table 3*).

From these discussions it should be clear that it is important not to allow the potential of the conducting salt to stray outside the stable region otherwise the electrode surface may become completely covered in insoluble decomposition products. For pressed pellet electrodes, these may be removed by polishing. For drop coated or cavity electrodes it will be necessary to make a fresh electrode surface if decomposition occurs. For single crystal electrodes, polishing is obviously impossible but it is possible to improve the situation by soaking the electrode in spectroscopic grade acetonitrile for 1 or 2 h.

Within the stable potential range TTF.TCNQ electrodes behave very much like conventional metallic electrodes for the oxidation of solution redox species such as ferricyanide (*Figure 13*).

5.2 Electrochemistry of NMP.TCNQ

The electrochemistry of NMP.TCNQ is very similar in outline to that for TTF.TCNQ. Within the stable range, and provided the electrode surface is clean so that there are no insoluble species present, the cyclic voltammogram is essentially featureless (*Figure 14*).

If the potential of the electrode is taken positive at around +0.5 V versus SCE the electrode material is oxidized to give insoluble TCNQ

$$NMP.TCNQ \rightarrow NMP^+ + TCNQ^\circ + e^-$$

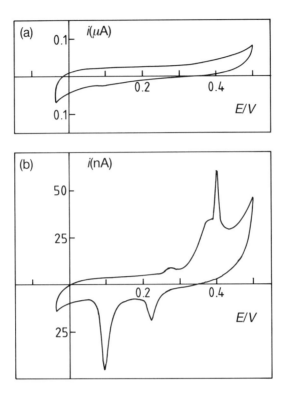

Figure 14. Cyclic voltammograms of single crystal NMP.TCNQ electrodes recorded at a potential sweep rate of 10 mV in phosphate buffer pH 7.2 containing 0.1 M tetraethylammonium tetrafluoroborate. (a) A clean electrode, (b) an electrode which has been cycled outside the stable potential range, the various peaks are caused by insoluble decomposition products.

The NMP^+ generated is generally soluble and diffuses away from the electrode.
On reduction of the electrode at potentials below -0.05 V versus SCE an insoluble reduced NMP^o species is formed

$$NMP.TCNQ + e^- \rightarrow NMP^o + TCNQ^{\cdot -}$$

As for the TTF.TCNQ case, the fate of the $TCNQ^{\cdot -}$ depends on the electrolyte. The exact nature of the insoluble neutral NMP species has not been established but it is electroactive at the electrode and gives rise to a peak in the cathodic sweep at about 0 V versus SCE presumably due to

$$NMP^o \rightarrow NMP^+ + e^-$$

which removes the material from the electrode surface.

Table 4. Stable potential ranges for conducting organic salts.

Conducting salt[a]	Anodic limit (V versus SCE)	Cathodic limit (V versus SCE)
TTF.TCNQ	+0.42	−0.23
NMP.TCNQ	+0.40	−0.24
Qn.TCNQ$_2$	+0.18	−0.22
Fc.TCNQ$_2$	+0.34	−0.22
Cu(dipy).TCNQ	+0.34	−0.22
TEA.TCNQ	+0.04	−0.11
TTT.TCNQ	+0.64	−0.23
Ac.TCNQ	+0.48	−0.22
Pb(DTF)$_2$	+0.37	−0.30

[a] Qn, quinoline; Fc, ferrocene; Cu(dipy), copper dipyridylamine; TEA$^+$, tetraethylammonium; TTT, tetrathiotetracene; Ac$^+$, acridinium; DTF, is 9-dicyanomethylene-2,4,7-trinitrofluorenone.

5.3 Other conducting organic salts

As we can see the usable potential region for any conducting organic salt electrode is determined by the decomposition of the material by electrochemical oxidation or reduction. Further depending upon the composition of the electrolyte the decomposition products may be insoluble and lead to new peaks in the cyclic voltammetry for the electrode within the stable region. The approximate stable potential ranges for a number of conducting salts are given in *Table 4*.

In conclusion, cyclic voltammetry is an excellent and very sensitive technique for checking the quality of conducting salt electrodes. Any electrodes which do not show well behaved cyclic voltammetric responses should be either repolished, repacked or discarded.

6. Enzyme electrodes

Amperometric enzyme electrodes can be made using the organic conducting salt as the working electrode material using either flavoproteins or NAD(P)H-dependent dehydrogenases (9). The enzymes can be used in either membrane electrodes or, in the case of some of the flavoproteins, by simply using the adsorbed enzyme. Examples of the different types of electrode are described below. The general principles are equally applicable to many other flavoproteins and dehydrogenases and it should prove relatively simple to extend these ideas to other systems.

6.1 Flavoproteins

The flavoproteins contain the flavin adenine dinucleotide (FAD) prosthetic group at the active site. The group is generally reasonably strongly bound, although it can be removed under certain circumstances. In the natural catalytic reaction, for example the oxidation of glucose catalysed by glucose oxidase, the FAD prosthetic group

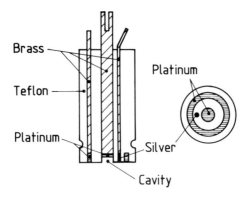

Figure 15. Construction of a self contained membrane electrode. The conducting organic salt is packed into the cavity and the concentric platinum ring acts as the counter electrode. A silver chloride coated silver wire is used as the reference electrode. In this design all three electrodes are behind the membrane.

is first reduced by reaction with substrate and then re-oxidized by reaction with oxygen. Thus the FAD cycles between its oxidized and reduced forms. This is easy to follow spectrophotometrically as the oxidized form of the enzyme is yellow and the reduced form colourless.

The overall catalytic reaction can be written in two steps:

$$GOD(FAD) + glucose \rightarrow GOD(FADH_2) + gluconolactone$$
$$GOD(FADH_2) + O_2 \rightarrow GOD(FAD) + H_2O_2$$

where $GOD(FAD)$ and $GOD(FADH_2)$ represent the oxidized and reduced forms of the enzyme respectively.

In the amperometric enzyme electrode we replace the second step, the oxidation by oxygen, with oxidation by the electrode. Thus for the amperometric enzyme electrode the reaction scheme becomes:

$$GOD(FAD) + glucose \rightarrow GOD(FADH_2) + gluconolactone$$

then at the conducting salt electrode

$$GOD(FADH_2) \rightarrow GOD(FAD) + 2H^+ + 2e^-$$

We therefore get two electrons for every molecule of glucose oxidized by our electrode. This is the same general principle for all the flavoprotein-based enzyme electrodes using organic conducting salts as the working electrode (10). Below we describe some examples of such electrodes.

6.1.1 Glucose electrodes

There are basically two types of glucose electrode; those with a membrane and those using adsorbed enzyme. We will consider the membrane electrode first.

Figure 15 shows a typical membrane electrode (11). Such electrodes can be made either with just the working electrode behind the membrane and using separate counter and reference electrode or, as in the figure, they can be self contained with all three electrodes behind the membrane. The latter configuration is often more convenient but does require a more sophisticated electrode construction.

The membrane is usually dialysis tubing (Visking) although other types of membrane can be used. The purpose of the membrane is to entrap the enzyme close to the surface of the electrode whilst still allowing the substrate, in this case glucose, to reach the electrode. Before use the membrane is prepared as follows:

Protocol 13. Preparation of dialysis membranes

1. Cut the dialysis membrane (cellulose acetate, Visking 36/32 relative molecular mass cut-off 12 000 – 14 000) into pieces large enough to cover the electrode.
2. Prepare a 1% w/v solution of sodium carbonate (Na_2CO_3) in purified water.
3. Place the pieces of dialysis membrane in the sodium carbonate solution and boil for 10 min.
4. Rinse the pieces of membrane with purified water. Store refrigerated in a solution of 0.01 M Tris buffer containing 1 M EDTA. Do not allow the membrane to dry out once it has been prepared.

The properties of the membrane can affect the response for the electrode so that in some circumstances it may be desirable to change the type of membrane in order to optimize the response of the sensor for a specific application. This will be discussed in Section 7.

Once the membrane and working electrode have been prepared it is a simple matter to assemble the enzyme electrode.

Protocol 14. Construction of a membrane electrode for glucose

1. Prepare the TTF.TCNQ working electrode and dialysis membrane as described in Sections 4.1 – 4.4 and the first protocol of Section 6.1.1. It is best to check the working electrode using cyclic voltammetry in background electrolyte before use.
2. Dissolve 16.5 mg of glucose oxidase (18 U mg^{-1}) in 1 ml of buffer.[a,b]
3. Select a piece of the prepared dialysis membrane and rinse it with purified water. Clamp the electrode and place one or two drops of enzyme solution on to its surface. Place the membrane over the electrode and secure with a rubber 'O'-ring. Stretch the membrane tight.

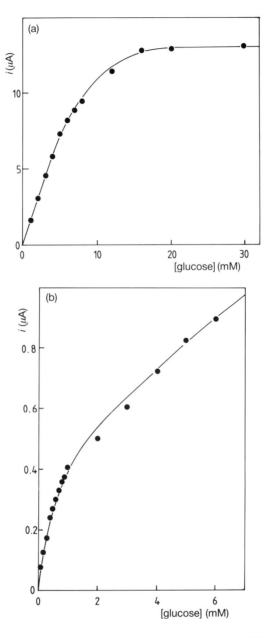

Figure 16. (a) Response to glucose for a membrane electrode based on TTF.TCNQ and glucose oxidase using a dialysis membrane. The electrolyte is 0.1 M NaCl, 0.15 M phosphate buffer at pH 7.2. (b) Results for the measurement of glucose using a TTF.TCNQ electrode with adsorbed glucose oxidase. The enzyme was adsorbed on the electrode by simply dipping the packed cavity electrode in glucose oxidase solution.

Protocol 14 continued

4. Trim off the excess membrane with a pair of scissors and wash the electrode and outside of the membrane copiously with purified water to remove any excess enzyme.

5. Transfer the electrode to the electrochemical cell containing the buffered background electrolyte and potentiostat the electrode at +0.05 V versus SCE.[c]

6. Allow the background current to decay to a steady low value (of the order of 5 μA cm^{-2} or less) before use. This can take around 30 min the first time the electrode is used but thereafter, as long as the same TTF.TCNQ electrode is employed, it should take no longer than a couple of minutes.

7. The electrode can now be calibrated using stock glucose solutions. The electrode should be kept in the same buffer solution as in step 2 at room temperature when not in use and should be recalibrated when next used.

[a] In general it is best to ensure as high an enzyme activity as possible behind the membrane so that the response is determined by transport through the membrane rather than by the enzyme kinetics. This is discussed in Section 7.

[b] The choice of buffer depends upon the particular circumstances but it is important to ensure that the buffer capacity is great enough to cope with the protons produced by the reaction and that there are sufficient ions present to carry the current. A suitable buffer is 0.1 M phosphate, 0.15 M NaCl, pH 7.2.

[c] The choice of potential is not too critical although it must lie within the stable range of the electrode. For TTF.TCNQ electrodes, potentials in the range 0.0 to +0.3 V versus SCE are suitable. It is best to select the potential that gives the lowest background current.

Essentially all that is necessary is to place a drop of enzyme solution behind the membrane and in contact with the electrode.

It is important to exclude bubbles and this may require a little practice. Once assembled the electrode should be kept in background buffer solution and allowed to stabilize before use.

Figure 16a shows typical results obtained for a membrane electrode for glucose using TTF.TCNQ as the working electrode material. Note that the response saturates at around 20 mM glucose for this electrode. The form of the response will be discussed in Section 7. Electrodes of this type can be very stable; electrodes based on TTF.TCNQ and glucose oxidase can be used for up to 28 days. The electrode can be stored in buffer solution at room temperature between measurements.

Although membrane electrodes are useful in many applications, there are situations where the use of a membrane is difficult or undesirable. In these circumstances adsorbed enzyme can be used. *Figure 16b* shows results for the detection of glucose using a TTF.TCNQ electrode without a membrane. In this case, the enzyme was adsorbed at the surface by dipping a packed cavity electrode in a solution of the enzyme for 15 min, followed by washing with purified water to remove the loosely bound material. The adsorbed enzyme is strongly bound to the electrode and such electrodes can be used for periods in excess of 1 day without problems.

The use of adsorbed enzyme is especially useful in the construction of small electrodes.

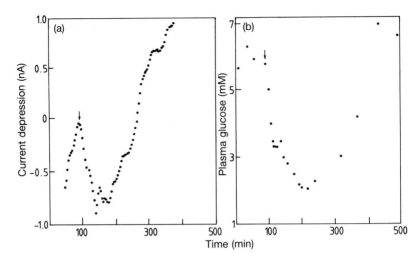

Figure 17. Comparison of the response from an implanted glucose electrode (a) with blood glucose as determined from blood samples (b). The arrow indicates the injection of 25 units kg^{-1} of insulin. Taken from ref. 17.

Protocol 15. Construction of mini-electrodes for implantation

1. Starting with a 4 cm length of 300 μm diameter Teflon-coated silver wire (AgIOT, Leico Industries Inc., USA), slide the Teflon insulation along the wire to create a 2 mm deep cavity. Stop further slip of the Teflon by gently squeezing flat a small length of 17 gauge hypodermic cannula around the wire.
2. Pack tightly the first 1 mm of the cavity with carbon paste, made from 0.7 g of carbon powder (Ultracarbon USP grade) in 0.25 ml of silicon oil (Aldrich high temperature grade), using a bare silver wire as a plunger. This prevents silver dissolution currents and provides good electrical contact.
3. Using small crystallites of TTF.TCNQ, tightly pack the remainder of the cavity.
4. Attach a small gold electrical contact to the other end of the wire.
5. Soak the electrode in 60 mg ml^{-1} glucose oxidase in a 50 mM acetate buffer solution pH 5.4 overnight to load with the enzyme.
6. Transfer to an electrochemical cell containing 0.15 M phosphate buffered to pH 7.4, deoxygenated and stirred by a stream of nitrogen. Potentiostat the electrode at 0.25 V versus Ag/AgCl for about 2.5 h.
7. Calibrate over the physiological range for glucose (0–10 mM) and ascorbic acid (0–0.5 mM).

Figure 17 shows results for a mini glucose electrode prepared in this way for *in vivo* studies in the rat brain (12). A useful technique with such electrodes to avoid

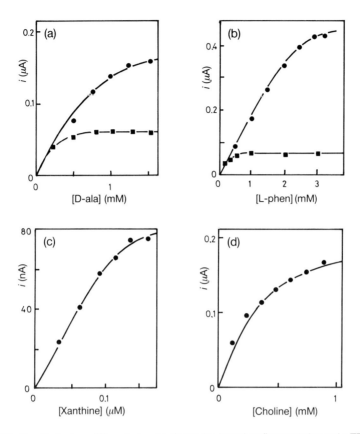

Figure 18. Results for four different enzyme electrodes based on flavoproteins and a TTF.TCNQ packed cavity electrode. The circles show data for membrane electrodes whilst the squares are for adsorbed enzyme. (a) D-amino acid oxidase, (b) L-amino acid oxidase, (c) xanthine oxidase, (d) choline oxidase.

problems with interference is to prepare two identical electrodes and then to boil one to deactivate the enzyme. The difference in the currents measured at the two electrodes, active and inactive, is then related to the glucose concentration. With this arrangement the effects of interfering species such as ascorbate are minimized.

6.1.2 Other substrates

Other flavoproteins can be used with TTF.TCNQ electrodes in exactly the same way as glucose oxidase. The resulting electrodes can then be used to measure a range of substrates. The flavoproteins which have been used successfully with TTF.TCNQ electrodes are glucose oxidase (EC 1.1.3.4), xanthine oxidase (EC 1.2.3.2), D-amino acid oxidase (EC 1.4.3.3), L-amino acid oxidase (EC 1.4.3.2) and choline oxidase (EC 1.1.3.17). This list is not exhaustive. *Figure 18* shows some typical results for

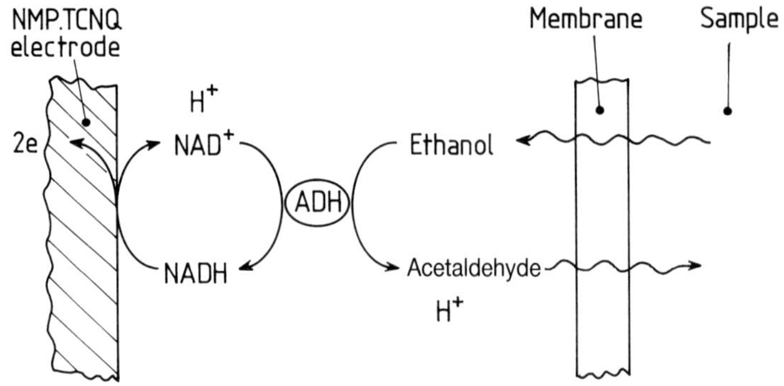

Figure 19. Operation of an amperometric enzyme electrode for ethanol. The ethanol diffuses through the membrane into the electrode where it is oxidized by NAD^+ to give NADH and acetaldehyde. The acetaldehyde diffuses out of the electrode and the NADH is reoxidized to NAD^+ at the NMP.TCNQ working electrode. The current is then related to the concentration of ethanol in the sample.

different flavoprotein-based amperometric electrodes constructed both with and without membranes.

6.2 NADH-dependent dehydrogenases

There are a large number (over 250) of NADH-dependent dehydrogenase enzymes. In these systems, unlike the flavoproteins, the NADH coenzyme is not strongly bound to the enzyme. A well studied example of this class is alcohol dehydrogenase, an enzyme which catalyses the reaction between NAD^+ and ethanol to give NADH and acetaldehyde:

$$CH_3CH_2OH + NAD^+ \xrightarrow{ADH} CH_3CHO + NADH + H^+$$

The reaction occurs by an ordered bisubstrate mechanism. In this mechanism the coenzyme first binds to the apoenzyme to give a complex referred to as the holoenzyme. The holoenzyme then binds the substrate and reaction occurs between the bound coenzyme and the bound substrate. This is followed by dissociation of the product and then dissociation of the reduced coenzyme to complete the catalytic cycle. The same mechanism applies for most NADH-dependent dehydrogenases.

In the amperometric enzyme electrode the NADH produced is reoxidized to NAD^+ at the working electrode:

$$NADH \rightarrow NAD^+ + H^+ + 2e^-$$

NMP.TCNQ is used as the electrode material for this purpose because the oxidation of NADH proceeds rapidly at around +0.1 V versus SCE. The current flowing at the working electrode is then related to the concentration of substrate present. This is shown schematically in *Figure 19*.

This system is potentially very flexible because of the large number of dehydrogenase enzymes which can be used, providing electrodes for a wide range of substrates. Some of the possibilities are alcohol, lactate, malate, glutamate, glucose, glycerol and 3-α-hydroxysteroids.

Many dehydrogenase enzymes are commonly used in spectrophotometric assays based on either the measurement of the initial rate of production of NADH or of the final amount of NADH produced. These are followed either by directly monitoring the NADH absorption or by coupling the system to a dye and following the production of a coloured dye species. Assays of this type can be converted into amperometric enzyme electrodes using the approach described below. The electrochemical assay has the advantage that it is generally much faster to carry out, requires less sample manipulation, and uses cheaper equipment. In addition the electrodes are more flexible; they can be used for continuous monitoring or in flow injection analysis.

One point which should be considered when designing amperometric enzyme electrodes based on dehydrogenase enzymes is that the reactions are often reversible. In other words there can be significant product inhibition. Spectrophotometric assay systems often overcome this by removing one of the products. This tips the balance in favour of the product side. The effects of product inhibition on electrode response are discussed in Section 7.

As examples of amperometric enzyme electrodes of this type we will describe dehydrogenase electrodes for ethanol and for bile acids. The general principles apply equally to the other NADH-dependent dehydrogenases.

6.2.1 Ethanol (13)

For the NADH-dependent dehydrogenases it is necessary to have both the enzyme and the coenzyme present at the electrode surface. This is achieved using a membrane to entrap the species. Exactly the same electrode geometry as used with the flavoproteins can be used for these systems.

Ideally the membrane should retain the coenzyme, NAD$^+$ with a molecular weight of 663, but allow the substrate and product to diffuse into and out of the electrode. In practice this is difficult to achieve satisfactorily because those membranes with a low molecular weight cut off slow down the diffusion of substrate so much that the electrode response becomes very sluggish (*Table 5*). For this reason ordinary dialysis membrane represents the best compromise at the present. However, it is clear that better membranes would be a great help in this area.

An alternative approach to this problem is to increase the molecular weight of the coenzyme. This can be achieved by the use of dextran modified NAD$^+$. This is commercially available (Sigma) and because of its high molecular weight is retained by ordinary dialysis membrane. However, the use of the dextran modified material reduces the sensitivity of the electrode to about 10% of that with ordinary NAD$^+$ and it is about 36 times the price! At present the simplest approach is to add NAD$^+$ to the buffer solution with the sample.

Table 5. Effect of membrane on electrode response[a].

Membrane	Molecular mass cut-off	Sensitivity (μA mM^{-1})	Response time (min)[b]
Ultrafiltration YCO5	500	2.3×10^{-2}	46
Millipore Duralon	—[c]	0.4	6
Benzoylated dialysis	2000	0.6	18
Ordinary dialysis	13 000	1.2	2

[a] Taken from ref. 14.
[b] Time to reach 95% of steady-state value.
[c] Pore size 1 μm.

The preparation of an ethanol electrode is given below.

Protocol 16. Enzyme electrode for ethanol

Preparing the electrode

1. Prepare the NMP.TCNQ working electrode and dialysis membrane as described in Sections 4.1–4.4 and the first protocol of Section 6.1.1.

2. Prepare a buffer solution containing 0.60 M Tris (Aldrich), 0.4 M lysine, and 0.017 M β-NAD$^+$ (Sigma) pH 9.[a]

3. Dissolve 10 units of yeast alcohol dehydrogenase [EC 1.1.1.1] in a few microlitres of the buffer.

4. Clamp the working electrode upright and place the enzyme solution on to the electrode.

5. Rinse the dialysis membrane with purified water and then place it over the electrode. Secure in place with a rubber 'O'-ring. Ensure that there are no bubbles trapped at the electrode surface. This requires a little practice.

6. Pull the membrane tight and trim away any excess. Wash the electrode copiously with purified water to remove any excess enzyme.

7. Place the electrode in the electrochemical cell containing a known volume of buffered NAD$^+$ solution and potentiostat the electrode at 0.0 V versus Ag/AgCl.

8. Allow the background current to decay to a steady value. Typically this will be between 0 and 3 μA and takes around 10 min.

9. The electrode is now ready to use.

Protocol 16 *continued*

Ethanol assay

10. Calibrate the electrode by injecting aliquots of a stock ethanol solution in buffer into a known volume of buffer in the electrochemical cell. After each addition allow the current to reach a steady-state value before measuring its value.
11. The same procedure is used on the unknown samples, diluting if necessary.
12. Replace the buffer solution after every three or four measurements.

a The lysine is present to remove the acetaldehyde as it is produced.

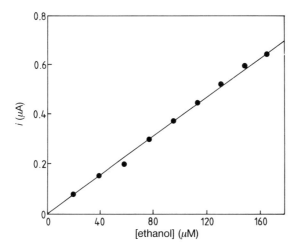

Figure 20. Typical calibration curve for an amperometric enzyme electrode for ethanol based on an NMP.TCNQ working electrode and using the enzyme alcohol dehydrogenase.

An alternative buffer solution is: 0.5 M glycine, pH 9.0 (Sigma) containing 0.05 M NaCl and 0.024 M semicarbazide to remove the acetaldehyde.

Once prepared the electrode can be used for periods up to 5 h using the same enzyme and membrane before degradation of the enzyme begins to affect the electrode response. After this time the electrode can be taken apart, washed and reassembled with a fresh enzyme solution ready to be used again. The above protocol also gives details of the assay procedure for ethanol using this type of electrode. Typical calibration curves are shown in *Figure 20*. Note that the response saturates at about 8 mM ethanol. This is determined by the saturated enzyme kinetics.

Finally we should note that the electrode also responds to higher alcohols such as propan-2-ol and butan-1-ol. This is not surprising as these are also substrates for the enzyme. In the absence of selective membranes enzyme electrodes can only be as selective as the enzyme used.

Conducting organic salt electrodes

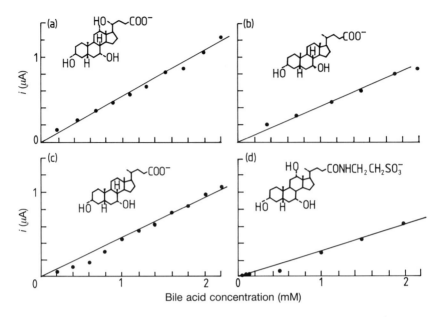

Figure 21. Results for the detection of four different bile acids obtained with a membrane electrode using 3-α-hydroxysteroid dehydrogenase. (a) Cholic acid, (b) ursodeoxycholic acid, (c) chenodeoxycholic acid, (d) taurocholic acid. Taken from ref. 15.

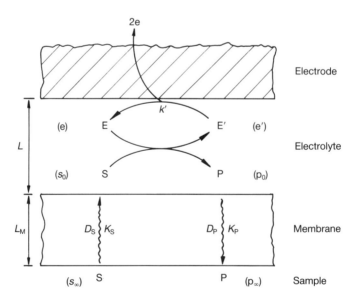

Figure 22. Kinetic scheme for the enzyme membrane electrode model. The meaning of the various symbols is explained in the text.

6.2.2 Bile acids

Another example of an NADH-dependent dehydrogenase which can be used in an enzyme electrode is 3-α-hydroxysteroid dehydrogenase (15). This is a useful system for the assay of bile acids and their taurine and glycine conjugates in samples of human bile. The conventional assay uses this enzyme and detects the amount of NADH generated when the reaction has gone to completion; this takes about 60 min In contrast to the spectrophotometric assay, the enzyme electrode assay takes just a couple of minutes for each sample.

Figure 21 shows calibration curves for four different bile acids, all containing a 3-α-hydroxy substituent. Over the range 0–2 mM the responses are linear.

7. The analysis of enzyme electrode data

The response of an amperometric enzyme electrode of the type described above, in which the enzyme is entrapped at the electrode surface behind a membrane, can be determined by any one of a number of kinetic steps (16). Fortunately it is possible to unravel these and to develop a relatively simple model for the electrode response. The great advantage of this is that it allows the rate limiting steps to be identified. This in turn means that rational steps can be taken to optimize or improve the electrode design and performance.

In this final section we shall present the model and describe its application to some real data.

7.1 The model

Figure 22 shows the scheme for a membrane enzyme electrode. The substrate, S, diffuses through the membrane to react with the entrapped enzyme in the thin electrolyte layer at the electrode surface. The product, P, is then lost by diffusion back through the membrane. Note that we must consider both the diffusion and partition of substrate and product in the membrane. These processes are characterized by the diffusion and partition coefficients, D_x and K_x, for the two species. Inside the electrolyte layer the reaction scheme is completed by reaction at the electrode characterized by the heterogeneous rate constant k'.

The model assumes that concentration polarization is negligible in the bulk solution; this corresponds to a well stirred sample. In addition, concentration polarization within the thin solution layer containing the enzyme is assumed to be minimal. This is a reasonably good approximation because this layer is only of the order of several μm in thickness.

If we begin by considering the flavoprotein case, the enzyme kinetics can be written as follows:

$$S + E_1 \underset{k_{-1}}{\overset{k_1}{\rightleftharpoons}} E_1S \underset{k_{-2}}{\overset{k_2}{\rightleftharpoons}} E_2P \underset{k_{-3}}{\overset{k_3}{\rightleftharpoons}} E_2 + P \qquad (1)$$

For each step in this scheme

$$K_n = k_n/k_{-n} \tag{2}$$

The overall equilibrium between $E_1 + S$ and $E_2 + P$ is described by

$$K_{TD} = K_1 K_2 K_3 \tag{3}$$

We must consider the transport of substrate and product through the membrane. This is done in terms of the mass transfer rate constants k_s' and k_p' where

$$k_x' = D_x K_x / L_M \tag{4}$$

and L_M is the membrane thickness.

This analysis leads to the following expression for the steady-state flux, j, in mol cm^{-2} sec^{-1}:

$$\frac{e_\Sigma}{j} = \left\{1 - \frac{j}{k_s' S_\infty}\right\} \left\{\frac{1}{Lk_{cat}} + \frac{1}{k'} + \frac{K_3^{-1}(1 + K_2^{-1})}{k'}\left[P_\infty + \frac{j}{k_p'}\right]\right\} + \frac{1}{S_\infty}\left\{\frac{K_M}{Lk_{cat}} + \frac{K_1^{-1} K_2^{-1} K_3^{-1}}{k'}\left[P_\infty + \frac{j}{k_p'}\right]\right\} + \frac{e_\Sigma}{k_s' S_\infty} \tag{5}$$

where

$$1/k_{cat} = 1/k_2 + 1/K_2 k_3 + 1/k_3 \tag{6}$$

and

$$K_M/k_{cat} = 1/k_1 + 1/K_1 k_2 + 1/K_1 K_2 k_3 \tag{7}$$

Each of the terms in equation 5 can be identified with a particular rate limiting process in the enzyme membrane electrode. Thus if we can fit our data to this model we can establish which step or steps are rate limiting.

This treatment can be extended to cover the NADH-dependent dehydrogenase enzyme electrode described above where the NADH shuttles between the enzyme and the electrode. In this case we assume that the kinetics of binding of the enzyme to NAD$^+$ are sufficiently rapid for equilibrium to be established between E and E.NAD$^+$:

$$E + NAD^+ \underset{}{\overset{K_0}{\rightleftharpoons}} E.NAD^+$$

The reaction scheme can then be written as follows:

$$S + E.NAD^+ \underset{k_{-1}}{\overset{k_1}{\rightleftharpoons}} S.E.NAD^+ \underset{k_{-2}}{\overset{k_2}{\rightleftharpoons}} P.E.NADH \underset{k_{-3}}{\overset{k_3}{\rightleftharpoons}} E + NADP + P + H^+$$

and at the electrode:

$$NADH \xrightarrow{k'} NAD^+ + H^+ + 2e$$

Under these circumstances, assuming that there is no product in the external solution, we obtain the following expression for the flux (9,13):

$$\frac{e_\Sigma}{j} = \frac{1}{Lk_{cat}}\left[1 - \frac{j}{k_s's_\infty}\right] + \frac{K_M}{Lk_{cat}s_\infty} + \frac{je_\Sigma}{k_p'k'K_{TD}s_\infty[NAD^+]} +$$

$$\frac{j^2}{Lk_p'k'K_{TD}s_\infty[NAD^+]}\left[\frac{1}{k_{-1}} + \frac{K_2}{k_{-1}} + \frac{1}{k_{-2}}\right] + \frac{e_\Sigma}{k_s's_\infty} \quad (8)$$

where now $K_{TD} = K_0 K_1 K_2 K_3$ and describes the overall position of equilibrium between $S + NAD^+$ and $P + NADH + H^+$.

7.2 Analysis of results

Equations 5 and 8 are cumbersome and difficult to work with. More useful relations can be derived by taking special cases of the general case. In particular if we take the situation where there is no product inhibition things simplify greatly and we obtain, after some algebra,

$$\frac{s_\infty}{j} = \frac{1}{k'_{ME}}\left[1 + \frac{s_\infty}{K_{ME}}\left\{1 - \frac{j}{k_s's_\infty}\right\}\right] \quad (9)$$

where

$$1/k_{ME}' = K_M/(e_\Sigma Lk_{cat}) + 1/k_s' \quad (10)$$

is the effective heterogeneous rate constant for the enzyme membrane electrode and

$$K_{ME} = [K_M(Lk_{cat})^{-1} + e_\Sigma (k_s')^{-1}]/[(Lk_{cat})^{-1} + (k')^{-1}] \quad (11)$$

K_{ME} is similar to the Michaelis constant in homogeneous enzyme kinetics. When the concentration of substrate is less than K_{ME} the system is unsaturated; the current is then proportional to the substrate concentration and is governed by the rate constant

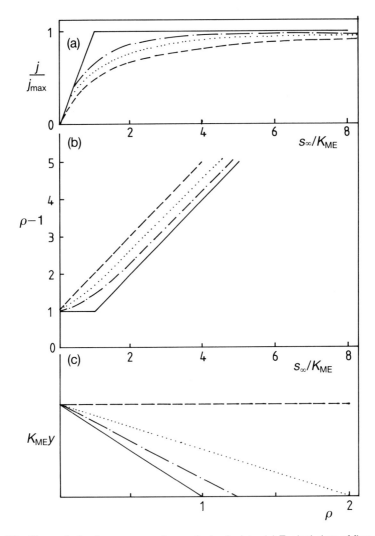

Figure 23. The analysis of enzyme membrane electrode data. (a) Typical plots of flux against substrate concentration for different values of k'_{ME}/k'_s under circumstances where there is no product inhibition. Curves (b) and (c) show the corresponding Hanes plots and plots of y respectively. The value of k'_{ME}/k'_s shown are as follows: —— 1.0; —·—· 0.8; · · · 0.5; - - - 0.0.

k'_{ME}. When the concentration of substrate is greater than K_{ME} the system becomes saturated and the current, as a function of substrate concentration, reaches a plateau.

Equation 9 is similar in form to the Hanes plot frequently used in the analysis of homogeneous enzyme kinetic data. Starting from equation 9 we can analyse our enzyme electrode data. This is described below.

Protocol 17. Analysis of membrane enzyme electrode data

1. Plot s_∞/i against s_∞, where s_∞ is the bulk substrate concentration and i the current (corrected for any background current). This plot may well be a curve.
2. Extrapolate the curve to estimate the intercept on the s_∞/i axis at $s_\infty = 0$. This gives $[s_\infty/i]_0 = 1/(nFAk_{ME}')$.
3. Using the data for values of s_∞/i significantly greater than $[s_\infty/i]_0$ calculate values of the dimensionless ratio: $\varrho = [i/s_\infty]/[i/s_\infty]_0$.
4. Calculate values of y where
$y = (\varrho^{-1} - 1)/s_\infty$.
5. Plot y against ϱ. This should give a straight line with a slope of $-k_{ME}'/(k_s' K_{ME})$, an intercept on the y-axis of $1/K_{ME}$, and an intercept on the ϱ axis of k'_s/k'_{ME}).
6. From the plots calculate K_{ME}, k_{ME}' and k_s'.

Figure 23 shows examples of calculated curves.

Figure 24 shows the various steps in the analysis of the data for a xanthine enzyme electrode. From the results in *Figure 24* we can see that for this electrode $K_{ME} = 0.15$ mM and $k_s'/k_{ME}' = 1.1$. In other words the diffusion of the substrate through the membrane is rate limiting. This is a desirable condition in many cases since it means that the response does not depend upon the activity of the enzyme.

Figure 25 shows the analysis for an ethanol electrode. In this case $K_{ME} = 6.3$ mM and $k_s'/k_{ME}' > 1$. Thus transport through the membrane is not the rate limiting process and for this electrode the response is limited by the enzyme kinetics. Consequently the response saturates above K_M for the enzyme and $K_{ME} \approx K_M$.

An alternative approach is to fit the electrode data to the appropriate equation using a non-linear least squares fitting routine. *Figure 26* shows data and best fit lines for a glucose electrode on the day it was prepared and after 65 h continuous operation in 30 mM glucose. It is clear that over this period the response has decreased by about 10%. However, the reason for this decrease is not immediately apparent. When we fit the data to the model we find that although the response changes it is still limited by diffusion of substrate through the membrane (*Table 6*). This illustrates the power of this type of analysis.

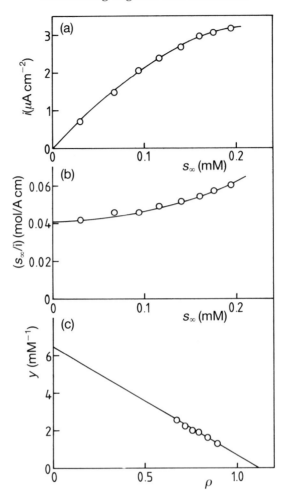

Figure 24. Analysis of data for a xanthine electrode based on xanthine oxidase and using TTF.TCNQ as the working electrode material. The analysis is carried out as described in Section 7.2. (a) Plot of the current as a function of concentration, (b) the equivalent of the Hanes plot for the data. Extrapolation of $s_\infty = 0$ gives an estimate of $[s_\infty/i]_0 = 1/nFAk_{ME}'$. (c) Plot of y against ϱ. The intercept on the y-axis gives the value for $1/K_{ME}$ and the intercept on the x-axis the value for k'_s/k'_{ME}. In this case the intercept on the x-axis is 1.1 indicating that diffusion through the membrane is rate limiting.

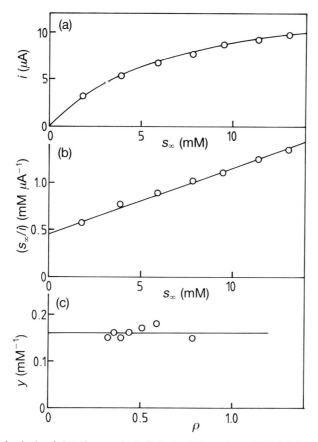

Figure 25. Analysis of data for an alcohol electrode based on alcohol dehydrogenase and using NMP.TCNQ as the working electrode material. The analysis is carried out as described in Section 7.2. (a) Plot of the current as a function of the concentration of ethanol, (b) the equivalent of a Hanes plot for the data, (c) the plot of y against ϱ. In this case the plot of y against ϱ is a horizontal line indicating that the enzyme kinetics are rate limiting for this electrode.

Table 6. Best fit parameters for glucose electrode data.

	Fresh	After 65 h
$nFAk_s'$ ($\mu A\ mM^{-1}$)	1.5 ± 0.3	1.1 ± 0.2
$nFALk_{cat}e_\Sigma/K_M$ ($\mu A\ mM^{-1}$)	6.7 ± 4.5[a]	6.7 ± 5.8[a]
$nFAe_\Sigma/\{(Lk_{cat})^{-1} + (k')^{-1}\}$ (μA)	13 ± 5	8.6 ± 3

[a] The errors are large for this parameter because the response is determined by the membrane transport term, $nFAk_s'$. Hence the enzyme kinetic term is poorly determined.

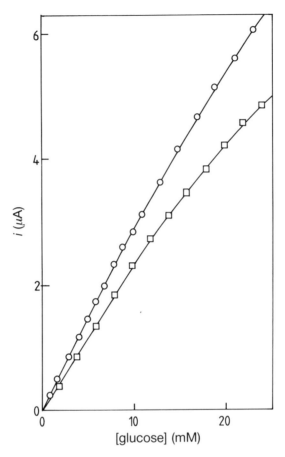

Figure 26. Data for a membrane electrode for glucose immediately after assembly (circles) and after 65 h continuous use in 30 mM glucose (squares). The lines are the non-linear least squares best fits to the membrane electrode model, equation 9. The fitting parameters are given in *Table 6*.

Acknowledgements

I would like to thank my colleagues upon whose work I have drawn and who kindly provided data for inclusion in this chapter. I am especially grateful to Lindy Murphy (Imperial College) for data on the paste electrodes, Drs Tony Cass and Vincent Sim (Imperial College) for ethanol electrode data, Dr Martyn Boutelle (Oxford) for the *in vivo* glucose data and finally Professor John Albery (Imperial College) in collaboration with whom much of this work originated.

References

1. Kulys, J. J. (1986). *Biosensors,* **2**, 3.
2. Fritchie, C. J. (1966). *Acta Crystallogr.,* **20**, 892.
3. Perlstein, J. H. (1977). *Angew. Chem. Int. Ed. Engl.,* **16**, 519.
4. Wheland, R. C. and Gillson, J. L. (1976). *J. Am. Chem. Soc.,* **98**, 3916.
5. Sawyer, D. T. and Roberts, J. L. (1974). *Experimental Electrochemistry for Chemists.* Wiley, New York.
6. Southampton Electrochemistry Group (1985). *Instrumental Methods in Electrochemistry.* Ellis Horwood, Chichester.
7. Jaeger, C. D. and Bard, A. J. (1980). *J. Am. Chem. Soc.,* **102**, 5435.
8. Jaeger, C. D. and Bard, A. J. (1979). *J. Am. Chem. Soc.,* **101**, 1690.
9. Albery, W. J. and Craston, D. H. (1987). In *Biosensors: Fundamentals and Applications.* Turner, A. P. F., Karube, I., and Wilson, G. S. (eds), Oxford University Press, Oxford, p. 180.
10. Albery, W. J., Bartlett, P. N., Bycroft, M., Craston, D. H., and Driscoll, B. J. (1987). *J. Electroanal. Chem.,* **218**, 119.
11. Albery, W. J., Bartlett, P. N., and Craston, D. H. (1985). *J. Electroanal. Chem.,* **194**, 223.
12. Boutelle, M. G., Stanford, C., Fillenz, M., Albery, W. J., and Bartlett, P. N. (1986). *Neurosci. Lett.,* **72**, 283.
13. Albery, W. J., Bartlett, P. N., Cass, A. E. G., and Sim, K. W. (1987) *J. Electroanal. Chem.,* **218**, 127.
14. Sim, K. W. (1990) *Biosensors,* in press.
15. Albery, W. J., Bartlett, P. N., and Cass, A. E. G. (1987). *Phil. Trans. R. Soc. Lond. B,* **316**, 107.
16. Albery, W. J. and Bartlett, P. N. (1985). *J. Electroanal. Chem.,* **194**, 211.

4

Immunoelectrodes

NICOLA C. FOULDS, JANE E. FREW, and MONIKA J. GREEN

1. Introduction

The range of analyte concentrations encountered in clinical chemistry is extremely large—from greater than 10^{-3} M for species such as glucose and cholesterol to less than 10^{-9} M for certain drugs and hormones. It is for the detection of these low level analytes that the application of immunological techniques is essential. In these circumstances not only is there a requirement for a sensitive assay; it must also be highly specific. At such low levels of analyte even trace amounts of interferents can become significant. The rationale for attempting to combine the fields of immunology and electrochemistry in the design of analytical devices is that such systems should be sensitive due to the characteristics of the electrochemical detector whilst exhibiting the specificity inherent in the antigen−antibody interaction.

The ideal situation would be to detect the binding of immunoreagents directly at an electrode, for example, by changes in surface potential. This could truly be described as an immunoelectrode. However, to date this has not proved successful and many promising early results have now been attributed to experimental artefacts that can be explained in terms of electrode kinetics (1). Instead what has emerged is an area of analytical technology referred to as 'electrochemical immunoassay' (2).

This chapter aims to describe the approaches being taken to design electrochemical immunoassays; more specifically amperometric immunoassays. After introducing the basic concepts of immunology, a brief discussion of relevant electrochemical techniques is included. Subsequent sections are designed to give an insight into the various strategies available for construction of analytical systems based on immunoassays with an electrochemical detection system. Details are given for the preparation of assay components and how to choose the best configuration depending on the nature of the substance(s) to be detected.

2. Basic concepts of immunoassays

2.1 The structure of antibodies

Antibodies belong to the family of glycoproteins known as immunoglobulins. They are synthesized by animals as part of the response to the presence of a foreign substance. Macromolecules capable of eliciting such a response are known as antigens

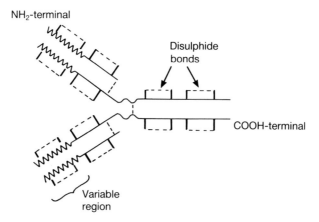

Figure 1. Diagrammatic representation of an IgG molecule showing arrangement of variable regions and disulphide bonds.

or immunogens. A hapten is a low molecular weight substance that is an antigen, but which is not, by itself, immunogenic. In order to elicit antibody formation the hapten must be bound to a macromolecule. The specificity of an antibody is directed against a particular site on an antigen known as the antigenic determinant or epitope.

Antibodies are heterogeneous with respect to physicochemical properties and function, but have similar basic structures. They consist of two kinds of polypeptide chains: heavy chains (50–70 kd) and light chains (~25 kd). Most immunoglobulins possess identical light chain–heavy chain pairs linked by disulphide bridges and non-covalent interactions. A Y-shaped structure results, with each half consisting of one light and one heavy chain linked together by a disulphide bond (*Figure 1*). The light and heavy chains are made up of a number of segments or domains, each of approximately 110 amino acids. The light chains contain two domains, whereas the heavy chains have 3–5 domains. It is the amino-terminal domains of both the light and heavy chains that are responsible for recognition and binding of the antigen.

Immunoglobulins may be divided into classes which are designated by the addition of a letter e.g. IgA, IgD, IgG. They differ in that although the heavy chains are of approximately the same size, they vary considerably in the amino acid sequence. On comparison of the amino acid sequence of the heavy chains of a particular class of immunoglobulin it is evident that approximately three quarters of each chain from the C-terminal end show very similar sequences—this is known as the constant region. The remainder of the peptide chain shows considerable variation in the amino acid sequence and is referred to as the variable region. It is the variations in amino acid sequence that give rise to the many different antigen binding sites.

An antibody will combine specifically with the corresponding antigen or hapten. The reaction of an antibody (Ab) with an antigen (Ag) to form a complex (Ab–Ag) may be represented as follows:

$$Ab + Ag \rightleftharpoons Ab-Ag$$

The intermolecular forces which contribute to the stabilization of the antibody−antigen complex are the same as those involved in the stabilization of the configuration of proteins and other macromolecules.

Antibody preparations may be polyclonal or monoclonal. A polyclonal antiserum will contain many different antibody molecules with varying affinities and specificities. Monoclonal antibodies, in contrast, are produced by a single clone of antibody-producing cells and therefore all molecules from the clone have the same specificity and affinity. The choice of which antibody preparation to use will depend upon the antigen, the application of the assay and the time and money available:

(a) If the antigen is not available in a pure form, as would be the case for certain cellular proteins, then a polyclonal antiserum cannot be obtained. However, for monoclonal antibodies the purity of the antigen is not as important due to the cloning and selection procedures used during their production.

(b) The use of polyclonal antibodies is limited by the fact that an antiserum taken from the same animal at different times differs in its properties. As a consequence, if an assay is to be based on the use of a polyclonal antibody it is necessary to have a sufficiently large pool of serum before beginning assay development. Monoclonal antibodies offer the advantage that once a suitable and stable clone has been found, virtually unlimited quantities of a particular antibody can be produced.

(c) The production of a monoclonal antibody with the correct characteristics for any given assay is expensive and time consuming. Polyclonal antibody production, however, is much cheaper and can be achieved in a shorter time.

2.2 Classification of immunoassays

An immunoassay may be defined as a technique based on the reaction between an antigen and an antibody for measuring the concentration of either reactant in solution. The method of classifying immunoassays is somewhat variable, but generally reflects the underlying principle and the nature of the label used to monitor the binding reaction. The first major distinction to be made is between homogeneous and heterogeneous immunoassays. A homogeneous system does not require separation of free and bound antigen; the assay relies upon the alteration of the electroactivity or function of the label on formation of the antibody−antigen complex. A more sensitive approach that is less prone to interference problems is the heterogeneous assay format in which there is a separation step. Within each of these classes there are a number of ways of designing the assay. The following configurations are relevant to the examples given in subsequent sections of this chapter:

(a) Competitive, solid-phase immunoassay in which antibody is immobilized on a solid phase and the antigen is labelled. In the test there is competition between the labelled and unlabelled antigen (i.e. the analyte) for the available antibody binding sites. After the binding reaction the amount of label associated with the solid phase will be inversely related to the concentration of analyte.

(b) Displacement immunoassay in which the antibody is again immobilized on a solid phase and the antigen is labelled. At the start of the assay all of the available binding sites on the immobilized antibodies are occupied by labelled antigen. On the addition of unlabelled antigen there is a displacement of labelled material and under appropriate conditions the extent of this displacement will be dependent upon the amount of analyte added.

(c) Sandwich immunoassay. This configuration is only suitable for use with high molecular weight antigens which possess at least two antigenic determinants or epitopes. A capture antibody is attached directly or indirectly to a solid phase. After the binding reaction a second labelled antibody against a different epitope is added in excess and the amount of labelled antibody associated with the solid phase is directly related to the amount of analyte.

There are very many types of label used to monitor the antibody–antigen binding reaction including: particles; metal and dye sols; radionuclides; enzymes; substrates and cofactors; electrochemically active compounds. Sections 4 and 5 present examples of electrochemical immunoassays using enzymes and electrochemically active species as labels.

2.3 Enzymes as labels in immunoassays

Enzymes are extremely useful as labels in immunoassays as their catalytic properties allow the detection and quantitation of low levels of immune reactants. The following are ideal properties for an enzyme to be employed in an enzyme immunoassay:

- high turnover number;
- low K_m for substrate, but a high K_m for product;
- stability upon storage (in the free or conjugated form);
- it must be possible to obtain a pure sample of the enzyme, i.e., it should be easy to prepare or be available commercially and should not be too expensive;
- enzyme activity should be easily detectable;
- absence of endogenous enzyme or interfering substances in the test sample;
- the conditions required for the use of the enzyme should be compatible with the assay conditions, e.g., in terms of pH, ionic strength, buffer composition.

No enzyme considered for use in an immunoassay will fulfill all the criteria listed above so a compromise has to be made. The enzymes most commonly used are alkaline phosphatase; horseradish peroxidase; glucose oxidase and β-galactosidase.

3. Electrochemical techniques for the development of amperometric immunoassays

The immunoassays described in Sections 4 and 5 involve the use of two electrochemical techniques known as DC cyclic voltammetry and chronoamperometry (3). The basic principles behind these methods are described in this section.

Conventionally they involve the use of a three-electrode system: a working electrode (i.e. the electrode at which the reaction of interest occurs); a reference electrode to maintain the potential of the working electrode, and a counter or auxiliary electrode to carry the cell current. In some circumstances, if small currents are involved (microamps or less) it is acceptable to use a two-electrode system (current can be drawn through the reference electrode without substantially affecting its potential). The electrochemical cell in which measurements are made is typically a glass, two-compartment cell in which the reference electrode is in a separate compartment to the working and counter electrodes. Connection between the two compartments is made via a Luggin capillary.

3.1 DC cyclic voltammetry

In voltammetry, a potential is applied to the working electrode and the current is measured. Cyclic voltammetry involves sweeping the potential between two limits, E_1 and E_2, at a fixed scan rate. On reaching the potential E_2 the scan direction is reversed whilst maintaining the same scan rate. At E_1 there are a number of alternatives: the potential scan may be (i) stopped; (ii) again reversed or (iii) continued to another potential value, E_3. The scan rates used in conventional experiments will vary from a few mV sec^{-1} to a few hundred V sec^{-1}. The use of higher scan rates is possible, but introduces experimental difficulties which restrict their application.

There are two components to the current measured at the working electrode. First, there is the capacitative or non-Faradaic component arising from redistribution of charged and polar species at the electrode surface. There is also the Faradaic contribution resulting from exchange of electrons between the electrode and redox active species either free in solution or immobilized at the electrode surface. If the potential is sufficiently oxidizing or reducing and the rate of electron transfer between the electrode and the redox species in solution is sufficiently fast, the Faradaic current is controlled by the rate of diffusion of the electroactive species to the electrode. For the reversible reduction of the redox active species O:

$$O + ne^- \rightleftharpoons R$$

with a redox potential E^o, the Faradaic current, i_f, will depend on the concentration gradient of O at the electrode surface according to the relationship:

$$i_f = nFAD_o \, (d[O]/dx)_{x=0}$$

where A is the area of the electrode, D_o is the diffusion coefficient of O and n is the number of electrons transferred.

If the working electrode is held at a positive potential and then the potential is swept towards and beyond E^o, the surface concentration of O will change according to the Nernst equation:

$$[O]/[R] = \exp\left[(E-E^o)nF/RT\right]$$

where E is the electrode potential, R is the gas constant and T is the temperature.

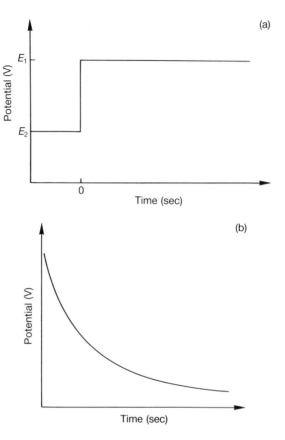

Figure 2. (a) Potential – time profile for a single-step chronoamperometric experiment. (b) Current – time response in a chronoamperometric experiment.

Reduction of O to R results in depletion of the concentration of O in the electrolyte layer close to the electrode surface (i.e. in the diffusion layer). The current is not maintained, but reaches a maximum value and then decays. On reversing the direction of the potential scan a peak arising from the re-oxidation of R will be obtained. For a reversible one-electron transfer reaction the following conditions apply:

- peak to peak separation of 57 mV;
- peak current proportional to (scan rate)$^{1/2}$ i.e. $i_p/\nu^{1/2}$ = constant;
- anodic and cathodic peak currents are identical.

In order to measure the Faradaic peak current from the cyclic voltammogram, the non-Faradaic component must be subtracted by drawing a tangent to the initial slope.

Cyclic voltammetry is a useful technique for an initial study of the electrochemical

3.2 Chronoamperometry

Chronoamperometry is a potential step technique in which the potential of the working electrode is changed instantaneously from initial to final value and the current–time response recorded.

Assuming the reaction of interest is the reduction of O and that initially only O is present in solution, a potential–time profile of the type shown in *Figure 2a* is applied to the working electrode. E_1 should be a potential at which neither reduction of O nor any other electrode reaction occurs. At time $t = 0$ the potential is instantaneously changed to a new value, E_2, at which the reduction of O occurs at a diffusion-controlled rate. A typical current–time profile for a chronoamperometric experiment is shown in *Figure 2b*. The form of this response can be understood by considering the concentration profiles for O and R during the course of the experiment. If the potential is stepped from a value significantly positive of the redox potential of the couple to one sufficiently negative, then at short times after the step the concentration of O will have changed from its initial value only at points very close to the electrode surface. The concentration profile will be very steep. Over the subsequent time period diffusion will cause the concentration profiles to relax towards their steady-state by extending into solution and becoming less steep. The current will decrease with time because it is a simple function of the flux of O at the electrode surface.

A discussion of the theoretical analysis of diffusion effects is provided in Chapter 9.

4. Alkaline phosphatase-labelled electrochemical immunoassays

4.1 Properties of alkaline phosphatases

Alkaline phosphatase [orthophosphoric-monoester phosphohydrolase (alkaline optimum) EC 3.1.3.1] is a broad and general term applied to non-specific phosphomonoesterases which exhibit optimum activity at alkaline pH. The hydrolysis of phosphomonoester in the presence of such an enzyme yields inorganic phosphate and the corresponding alcohol, phenol, etc:

$$R-OPO_3^{2-} + H_2O \rightarrow ROH + HPO_4^{2-}$$

A similar group of enzymes, the acid phosphatases, differ primarily in that they require acidic pH for their catalytic activity.

Many different sources of alkaline phosphatase exists, the most common being *Escherichia coli*, fungi and a variety of mammalian sources (intestine, bone, placenta, kidney, liver, lung and spleen). Alkaline phosphatases are also found in the Indian leech, Drosophila, the surf clam and some fish, but are absent from higher plants.

Immunoelectrodes

The enzymes have molecular weights in the range 80–150 kd and contain at least two atoms of zinc per molecule. Magnesium is also considered essential for maximum enzymatic activity.

Many chemical compounds inhibit alkaline phosphatases. Some are isozyme-specific, but two that are fairly universal inhibitors and are particularly relevant to the present discussion are ethylenediamine tetraacetic acid (EDTA) and inorganic phosphate. EDTA is commonly used in the preparation of dialysis membranes and phosphate is a popular choice of buffering medium in many experimental procedures. Both should be avoided when working with alkaline phosphatases.

4.2 Measurement of alkaline phosphatase activity

The activity of alkaline phosphatase enzymes is extremely sensitive to experimental conditions such as pH, ionic strength, buffer type and substrate concentration (4). An assay procedure for measuring alkaline phosphatase activity is described below.

Protocol 1. Determination of alkaline phosphatase activity

1. Immediately prior to the experiment prepare a solution of 5 mM p-nitrophenyl phosphate in 1 M diethanolamine buffer, pH 9.8 containing 0.5 mM $MgCl_2$.

2. Ensure the cuvette holder of the spectrophotometer is thermostatted to 37°C; adjust the wavelength to 410 nm and reference (blank) the spectrophotometer with the 5 mM p-nitrophenyl phosphate solution.

3. Pipette 2.9 ml of p-nitrophenyl phosphate solution into a 3 ml cuvette and ensure that it is equilibrated to 37°C.

4. Add 100 μl of a stock enzyme solution to the 3 ml cuvette (when dealing with calf intestinal alkaline phosphatase a 1 nM stock enzyme solution may be used). Mix the contents of the cuvette thoroughly and then measure the rate of change of absorbance (ΔA) over the initial 2 mins.

5. The enzyme activity in units (where 1 unit is defined as the amount of enzyme that liberates 1 μmol of p-nitrophenol from p-nitrophenyl phosphate per minute under the given conditions) can be calculated from the Beer–Lambert Law which relates absorbance to the concentration of the absorbing species:

$$\text{absorbance} = \epsilon\, c\, l$$

where ϵ is a constant known as the molar extinction coefficient; c is concentration and l is the pathlength of the cuvette. For p-nitrophenol $\epsilon = 18.8 \times 10^3\ M^{-1}\ cm^{-1}$. Assuming use of a standard cell of pathlength 1 cm:

$$[p\text{-nitrophenol}] = \frac{\Delta A}{18.8 \times 10^3 \times 2}\ \text{mol}\ l^{-1}\ \text{min}^{-1}$$

$$= \frac{\Delta A \times 10^6}{18.8 \times 10^3 \times 2}\ \mu\text{mol}\ l^{-1}\ \text{min}^{-1}$$

Protocol 1 continued

$$\text{i.e. } [p\text{-nitrophenol}] = \frac{\Delta A \times 3}{37.6} \; \mu\text{mol min}^{-1}$$

A transphosphorylating buffer, diethanolamine, is used as it is an extremely efficient phosphate acceptor and supports very high enzyme activities. When purchasing alkaline phosphatase it is essential to ascertain the assay conditions to which the enzyme units refer—even a change of buffer type may enhance the observed activity by a factor of three.

4.3 Alkaline phosphatase as an enzyme label

Alkaline phosphatases are used extensively as labels in immunoassays. The enzyme from calf intestine is particularly well suited to this role as it satisfies many of the criteria listed in Section 2.3. It is readily available; chemical modification of the enzyme is both easy and results in no loss of catalytic activity; it is also the most active of all alkaline phosphatases, having an activity in excess of 2000 U mg^{-1} under optimum assay conditions. Calf intestinal alkaline phosphatase has a very broad specificity range; hydrolysis has been observed for compounds containing P−F, P−O−C, P−O−P, P−S and P−N bonds. As already mentioned, an important consideration when choosing an enzyme label is the affinity of the enzyme for its substrate. It is useful if the K_m for the substrate is sufficiently low and/or solubility of the substrate is sufficiently high such that zero order kinetics can be achieved in the assay—alkaline phosphatases exhibit K_m values of the order of 1 mM for most of their substrates.

4.4 Electrochemically active substrates for alkaline phosphatase

A number of factors must be taken into account when assessing the suitability of an enzyme substrate for use with an electrochemical detection system. These may be summarized as follows:

- the electrochemistry of the substrate
- the electrochemistry of the product of the enzymatic reaction
- the medium in which the measurements will be performed
- the electrochemistry of endogenous materials in the test sample(s)

Ideally, the action of the enzyme on the substrate should result in a large change in the observed electrochemistry of the system. The potential limits between which these changes must occur are determined by the medium in which the measurements are to be performed (5). In an aqueous system the relevant potential window is approximately −0.25 V to +0.9 V versus SCE, i.e. between the potentials for the reduction of dissolved oxygen and the oxidation of water, respectively. The possible electroactive nature of endogenous substances in test samples must also be taken

Immunoelectrodes

into account. For example, in blood, compounds such as uric acid and ascorbic acid can be easily oxidized at potentials anodic of +0.35 V versus SCE. This means that for measurements in blood the substrate and/or product must be electroactive in the potential range −0.25 V to +0.35 V versus SCE.

The evaluation of an electroactive enzyme substrate or product may be achieved using DC cyclic voltammetry (see Section 3.1). Since the observed electrochemistry for a given system may be dependent upon factors such as pH, electrode material etc. it is important to examine the test system under conditions that are pertinent to the final application. Several electrochemically active substrates for alkaline phosphatase have been reported in the literature e.g. phenyl phosphate and naphthyl phosphate. Phenyl phosphate has been used extensively as a label in immunoassays (6). It has no electrochemistry within the potential window for aqueous solutions, but its hydrolysis product, phenol, can be oxidized at potentials around +0.8 V versus SCE. For a well defined test system such as an aqueous buffer, it is quite acceptable to measure an increasing signal arising from the dephosphorylation of phenyl phosphate, at such a positive potential. However, if the test were to be performed in blood or serum a large proportion of the signal would be due to the oxidation of endogenous sample components. An alternative substrate for alkaline phosphatase which avoids this problem is *p*-aminophenyl phosphate. This phosphate ester exhibits an irreversible wave in cyclic voltammetry at around +0.45 V versus Ag/AgCl, but its hydrolysis product *p*-aminophenol has reversible electrochemistry with a half-wave potential of −0.065 V versus Ag/AgCl (*Figure 3*). Product formation can therefore be monitored at potentials of the order of +0.1 V with no interference from the parent phosphate ester or from endogenous sample components. *p*-Aminophenyl phosphate may be prepared by reduction of *p*-nitrophenyl phosphate.

Protocol 2. Synthesis of *p*-aminophenyl phosphate

1. Dissolve 42 g of *p*-nitrophenyldihydrogen phosphate in 100 ml water and add a 10% (w/v) sodium hydroxide solution to give a final pH of 9.
2. Add 91.2 g of sodium sulphide nonahydrate and heat to 90−95°C in an oil bath for 1 h.
3. Cool the solution and then strongly acidify with concentrated hydrochloric acid.
4. Filter the acidified solution and adjust the pH of the acid filtrate to pH 4−5 using 25% sodium hydroxide. Allow the solution to cool.
5. Collect the white crystals that form; these will be the monosodium salt of *p*-aminophenyl phosphate.
6. Recrystallize from boiling methanol to give a white solid with a melting point of 181−183°C.

4.5 Hapten conjugation

4.5.1 Preparation of macromolecule – hapten conjugates

It is not the purpose of this chapter to give an extensive review of methods available

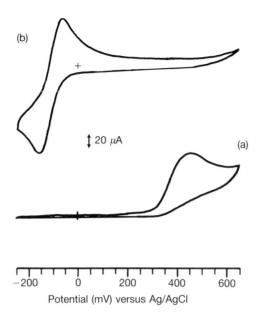

Figure 3. DC cyclic voltammograms of p-aminophenyl phosphate (a) before and (b) after addition of alkaline phosphatase. Experimental conditions: 0.1 M diethanolamine buffer, pH 9.5 containing 100 mM NaCl and 1 mM $MgCl_2$. Scan rate 50 mV s^{-1}.

for conjugation of haptens to macromolecules, but rather to note a few generalities in the procedures and describe a method that we have found very useful for preparing alkaline phosphatase–hapten conjugates [a more extensive discussion of this subject may be found in the text by Tijssen (7)].

Haptens contain a wide variety of reactive groups which provide useful chemical sites for conjugation procedures. The decision as to which linker positions on the hapten should be used in conjugation will be dictated by factors such as which closely related compounds could be encountered in the test sample and of course, whether a highly selective or a generic test is required. It is important to realize that the reactive groups may also be specific immunodeterminants, distinguishing the hapten from other closely related species. Conjugation through these groups will decrease the specificity of anti-hapten antibody. For instance, if a highly specific assay for theophylline is required and caffeine is likely to be encountered in the test samples then it would be inadvisable to use the 7 position of the xanthine as the point of linkage since this is the position that contains the only methyl group distinguishing the two molecules (*Figure 4*). Superior specificity may be achieved by choosing a group common to related molecules as the site of linkage. The use of a spacer between the hapten and the protein may also improve specificity and recognition of the hapten.

In practice, the choice of linker position may be limited by the chemistries of the groups involved. Also, better results will generally be obtained if the hapten is

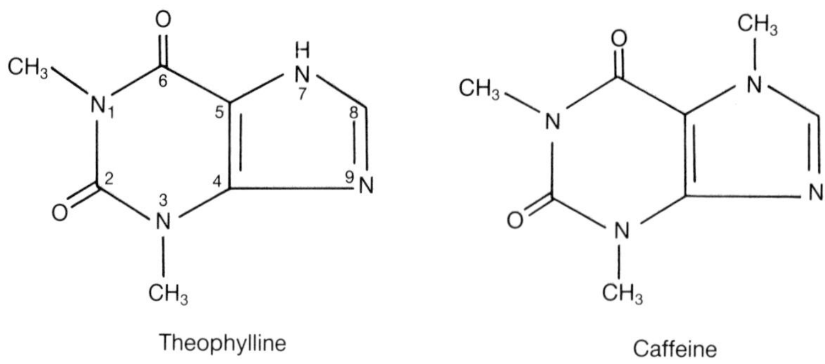

Figure 4. Structures of theophylline and caffeine.

conjugated to the enzyme via the same position as that which was used to prepare the immunogen—the antibody will be directed primarily against that part of the hapten furthest removed from the linkage to the carrier protein.

The extent of haptenation of the carrier protein molecule is an important parameter. For instance, to make a BSA–hapten immunogen it has been noted that an optimum effect is achieved with between 8 and 25 hapten molecules per protein molecule. For an enzyme-labelled hapten, both the range over which the hapten needs to be detected and the overall assay format will affect the degree of substitution that is optimal.

The linkage of haptens to proteins generally occurs at the most reactive residues on the proteins i.e. ϵ- and α-amino, phenolic, sulphydryl, imidazole and carboxyl groups. The reactivity of the various sites and thus, the point at which linkage is achieved, will be dependent upon the pH of the reaction mixture. The unprotonated forms of the nucleophilic groups are reactive and the degree of protonation is dependent upon the pK_a's of the groups. The microenvironment of the residue also governs reactivity.

Many conjugation procedures have been reported in the literature. A protocol known as the mixed anhydride method has proved particularly popular and is relatively easy to perform. It can be used with haptens that contain carboxyl groups or which can be carboxylated. The carboxyl group of the hapten is converted to an acid anhydride which can then react with the amino groups of the protein. A detailed procedure for preparation of a phenytoin–alkaline phosphatase conjugate via phenytoin acetic acid is given below.

Protocol 3. Synthesis of 5,5'-diphenylhydantoin-3-acetic acid

1. Dissolve 2.74 g of phenytoin (sodium salt; Sigma) in 50 ml of anhydrous dimethylformamide (Aldrich-SureSeal) at room temperature.
2. Add 1.79 g of methyl bromoacetate (Aldrich) and stir the reaction for 30 min. The temperature of the solution may rise to about 30°C and the mixture should

Protocol 3 continued

   become clear. Add 450 ml of water to the solution and leave the resulting suspension for 10 min.

3. Extract the resulting fractions exhaustively with ethyl acetate and dry the collected fractions by adding magnesium sulphate.

4. Filter the solution to remove the inorganic residues and then evaporate down to a solid.

5. Recrystallize the solid from methanol to give a white material melting at 185°C. This is 5,5'-diphenylhydantoin-3-methylacetate.

6. Hydrolyse the methyl ester to the acid by dissolving it in toluene and then adding 0.5 M sodium hydroxide solution (10% by vol.).

7. Reflux the mixture for 90 min; leave it to cool and then acidify until the pH is less than 1 and extract the phenytoin acetic acid into ethyl acetate.

8. Wash the organic fraction with water to remove traces of sodium chloride and then dry it over magnesium sulphate.

9. Filter the solution and evaporate to dryness. Recrystallize the resulting material from ethanol to give a white solid.

Protocol 4. Preparation of a phenytoin – alkaline phosphatase conjugate using the mixed anhydride method

1. Dissolve 2 mg of 5,5'-diphenylhydantoin-3-acetic acid in 100 µl of dry dimethylformamide (Aldrich-Sureseal) and cool to 0°C.

2. Add 5 µl of N-methylmorpholine followed by 5 µl of isobutylchloroformate. Leave the mixture stirring for 10 min at 0°C.

3. Remove 50 µl of the above reaction mixture and add it dropwise, with stirring, to a solution of 1.4 mg of alkaline phosphatase (Boehringer Mannheim immunoassay grade from calf intestine) in 10 ml of 25 mM tetraborate buffer, pH 9.2. Leave the reaction stirring at room temperature for 30 min. At first the solution may be opaque, but it should clear within the 30 min reaction period.

4. Prepare a suitable length of dialysis tubing by boiling in purified water for 10 min (do not use EDTA as this is a potent inhibitor of alkaline phosphatase) and dialyse the conjugate against 4 litres of 30 mM Tris buffer, pH 7.6 containing 1 mM $MgCl_2$, 1 M NaCl and 0.1 mM $ZnCl_2$. The dialysis buffer should be changed five times over a period of 48 h.

5. Store the conjugate in the presence of 1% bovine serum albumin and 15 mM sodium azide.

There are two additional experimental details worth noting:

(a) Boehringer Mannheim supply a preparation of alkaline phosphatase specifically for use as an enzyme label in enzyme immunoassay. Alternative enzyme preparations may require extensive dialysis before use;

(b) for successful conjugation with the mixed anhydride method all components used in the activation procedure must be dry. Suitable precautions include freshly distilling solvents before use or purchasing Aldrich 'Sure Seal' reagents; keeping all reagents in a desiccator and ensuring that all glassware is placed in a hot oven for approximately 30 min before use and then allowed to cool in a desiccator.

4.5.2 Characterization of enzyme – hapten conjugates

After completion of a conjugation procedure it is important to check that a conjugate has indeed been formed. A suitable method for such a screen would be an ELISA procedure (8) using a microtitre plate-bound hapten-specific antibody. Free hapten in the relevant concentration range is competed with various dilutions of the conjugate for specific antibody sites.

If the conjugate is to be used in an enzyme immunoassay it should be characterized in terms of the specific and total activity of the enzyme recovered and the degree of substitution (i.e. the number of hapten molecules per protein molecule). For an alkaline phosphatase conjugate the enzyme assay described in *Protocol 1* can be used to assess the activity of the conjugate. The figure obtained should then be related to the number of milligrams of protein to give a specific activity which can be compared with that of the original enzyme preparation. In many cases, increasing substitution will decrease the enzyme's specific activity.

The number of hapten molecules per protein molecule must be established as this ratio will affect the avidity of the antibody for the conjugate. There are a number of methods available and a choice should be made according to the physical properties of the hapten. A spectrophotometric method can be used if the hapten has a strong absorbance in the UV/visible region of the spectrum that is remote from the protein absorption. Other techniques that are more generally applicable include determination of the number of free amino groups on the protein before and after conjugation (e.g. by the colour development due to the reaction of 2,4,6-trinitrobenzene sulphonic acid with ϵ-amino groups) and the use of radioactive haptens.

Another test which provides some insight into the possible sensitivity of an assay using the prepared conjugate involves the determination of the minimum concentration of conjugate that can be detected. The experiment should be performed using equipment and conditions that are as close as possible to those that will be employed in the final immunoassay. *Figure 5* shows the results of a chronoamperometric experiment which involved the detection of *p*-aminophenol formed by the hydrolysis of *p*-aminophenyl phosphate in the presence of a phenytoin – alkaline phosphatase conjugate. Diluted samples of the conjugate were added to a fixed volume of 10 mM *p*-aminophenyl phosphate. After a given period the extent of formation of *p*-aminophenol was established by electrochemical interrogation of the solution using a carbon-based working electrode held at +50 mV versus Ag/AgCl. The data shown relate to the observed currents 45 sec after stepping the potential. The results demonstrate the sensitivity of the system—20 pM conjugate can be detected with a 2 min incubation.

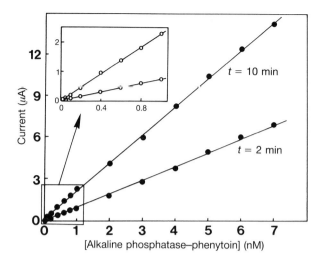

Figure 5. Detection of low levels of alkaline phosphatase conjugates.

4.6 Antibody purification

Antibody preparations are supplied in varying degrees of purity. A crude sample should be purified to some extent before proceeding further. Immunoglobulins have solubility characteristics unlike those of most other serum proteins and therefore a useful first stage in any purification is one in which separation is based on physicochemical properties. For polyclonal antisera this is commonly a salting-out step such as that decribed below.

Protocol 5. Salting-out procedure for polyclonal antisera

1. Place a known volume of antiserum in a glass test tube and whilst shaking the tube gently, gradually add a volume of saturated ammonium sulphate solution such that the final solution is 30% saturated. A white precipitate of immunoglobulin will begin to form.
2. Leave the solution to stand for 30 min.
3. Discard the supernatant and add a volume of 30% ammonium sulphate solution (diluted in water) so that the overall volume is the same as the original volume of antiserum. Resuspend the pellet in the ammonium sulphate and leave the suspension to stand for 60 min.
4. Centrifuge the suspension at 3000 r.p.m. for 30 min.
5. Discard the supernatant and dissolve the pellet in twice the original volume of 0.1 M phosphate buffer, pH 8.2 containing 150 mM sodium chloride.
6. Dialyse the immunoglobulin solution against 0.1 M phosphate buffer, pH 8.2

Protocol 5 continued

containing 150 mM sodium chloride for 24 h with one change of dialysis buffer over this period.

7. Store the preparation in the presence of 15 mM sodium azide (an antibacterial agent).

It is a cheap and efficient procedure which avoids any direct effects on the proteins and concentrates the purified protein. The method is based on the fact that IgG can be precipitated by the addition of ammonium sulphate to a final concentration of 30% of the saturation level, whereas other contaminating proteins will not precipitate at this salt concentration. The addition of ammonium sulphate, whether in the form of the crystalline solid or as a saturated solution, must be carried out gradually with continual stirring to avoid very high local salt concentrations which will cause precipitation of contaminants. Reduction of the final salt concentration will yield immunoglobulin of higher purity, whilst increasing it leads to higher yields with elevated levels of contaminating proteins. Further purification steps are often employed including affinity chromatography (9) and the use of ion-exchange materials (7). A disadvantage of the latter procedure is that in a polyclonal antiserum the highest affinity antibodies may be lost.

4.7 Solid-phase immobilization of antibodies

Immobilization of hapten-specific antibodies on a solid phase is a very popular way of effecting separation of bound from unbound conjugate and many different materials and types of surfaces have been examined for this purpose. In general, the solid phase should exhibit the following characteristics:

- high capacity for binding immunoreactants (high surface/volume ratio)
- offer the possibility for immobilization of many different immunoreactants
- cause negligible denaturation of the immobilized molecule
- allow orientation of immobilized antibody with the binding sites towards the solution

Various forms of plastic are the most popular solid-phase materials because immobilization procedures are then very simple. Despite its widespread use, the mechanism by which adsorption of antibodies on to plastic occurs is not well understood. In general, it is attributed to non-specific hydrophobic interactions and is independent of the net charge on the protein. The procedure has several disadvantages:

- the avidity of the antibody for the compound of interest may be significantly reduced, particularly in the case of larger antigens
- immobilization of the antibody reduces the rate of formation of the immune complex—the rate of antigen–antibody interactions is slower than in solution or with a particulate solid phase since only the antigen will be freely diffusing
- adsorption of antibodies to plastic surfaces may not always be reproducible

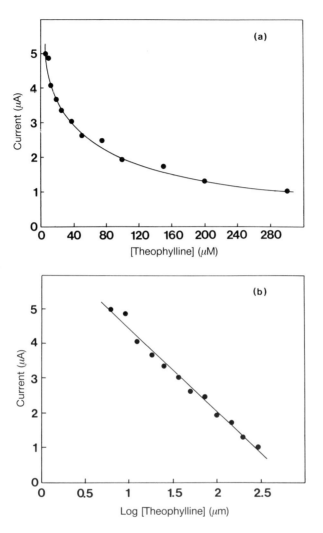

Figure 6. An electrochemical immunoassay for theophylline using Immunodyne membranes as the solid phase.

Immobilization on pre-activated macroporous nylon membranes such as Immunodyne (supplied by Pall) is more reproducible and is not difficult to achieve; simple protocols are supplied with the membranes. This support is advantageous both in terms of the covalent nature of the protein–membrane interaction and the fact that the membrane offers a high surface area and hence a high binding capacity. It is possible to bind up to 100–200 µg of protein per square centimetre of Immunodyne.

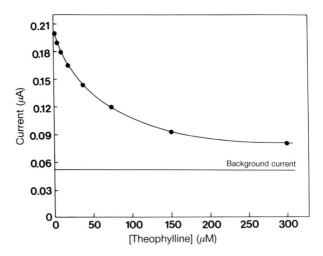

Figure 7. An electrochemical immunoassay for theophylline using a microtitre plate as the solid phase.

4.8 Configuration of electrochemical immunoassays

The following sections describe the use of alkaline phosphatase—antigen conjugates in competition and displacement assays with electrochemical detection of the enzyme label. Theophylline and phenytoin are used as model analytes.

4.8.1 A competition assay

In the case of an assay for theophylline, an antibody specific to theophylline is immobilized on an immunoaffinity membrane. The assay is based on competition between sample drug and drug—alkaline phosphatase conjugate for a fixed number of antibody binding sites. After an appropriate reaction time the membrane is exposed to p-aminophenyl phosphate. The subsequent formation of p-aminophenol occurs to an extent that depends upon the amount of conjugate present on the membrane. The greater the concentration of drug in the sample, the smaller the quantity of conjugate that will be able to bind to the membrane and hence, the lower the amount of p-aminophenol produced in any given time period. *Figure 6a* and *b* shows the results obtained over the clinically relevant range of theophylline concentrations. The membrane was incubated with p-aminophenyl phosphate for 3 min and the formation of p-aminophenol was monitored by its oxidation at a carbon electrode poised at +50 mV versus Ag/AgCl. It is interesting to compare these results with those shown in *Figure 7* where the solid phase for immobilization was a microtitre plate. For both systems maximum binding to the solid phase was achieved using the same antibody preparation and the same geometric area for immobilization. All other reagents and conditions were identical except that in *Figure 7* a 45-min incubation of bound conjugate with substrate was required to achieve a maximum

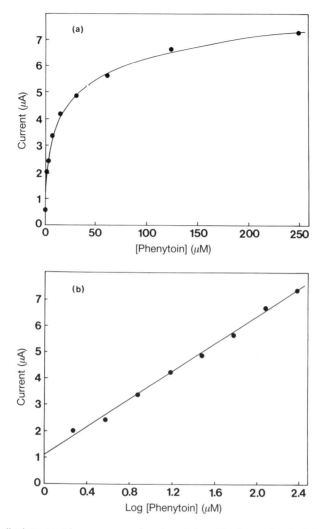

Figure 8. A displacement immunoassay for phenytoin with electrochemical detection.

current of approximately 0.1 µA whereas in *Figure 6a* the maximum current of 6 µA was achieved after only a 3-min incubation. These differences are due solely to the high binding capacities of the macroporous membrane.

4.8.2 A displacement assay

The development of a displacement assay is illustrated by way of reference to an assay for phenytoin. An antibody specific to phenytoin is immobilized on a macroporous membrane. The membrane is then exposed to phenytoin–alkaline

Table 1. Physicochemical properties of glucose oxidase from *Aspergillus niger*.

Molecular weight (kd)	186
Number of subunits	2
Coenzyme	FAD (2 molecules)
Overall shape	Globular
Carbohydrate content	16%
Isoelectric point	4.5
Extinction coefficient at 280 nm (1 mg ml^{-1} soln; 1 cm path length)	1.8

phosphatase conjugate at a concentration that will give maximum loading. After thoroughly washing the membrane to remove unbound conjugate, small disks are punched out and it is on these disks that the displacement assay is performed. The disks are incubated with various concentrations of phenytoin for an appropriate time period. The displaced conjugate is then assayed by detection of its alkaline phosphatase activity. A representative calibration curve for phenytoin over the clinical range is shown in *Figure 8a* and *b*.

5. Glucose oxidase in electrochemical immunoassays

5.1 Properties of glucose oxidase

Glucose oxidase (β-D-Glucose:oxygen 1-oxidoreductase, EC 1.1.3.4) is a flavin-containing glycoprotein which catalyses the oxidation of glucose to gluconolactone:

$$\text{Glucose} + O_2 \rightarrow \text{Gluconolactone} + H_2O_2$$

It can be obtained from a number of fungal sources, most commonly *Aspergillus niger* and *Penicillium* species. The enzyme from *A. niger* tends to be used in enzyme immunoassay and some of its physicochemical properties are listed in *Table 1*. Despite the relative abundance of the enzyme, limited structural information is available and a crystal structure has yet to be determined. The native enzyme contains two molecules of flavin adenine dinucleotide (FAD) and it is this flavin cofactor that is responsible for the redox properties of the enzyme. Structural information suggests that the FAD is firmly bound, but not covalently linked, to the polypeptide portion of the enzyme (denaturation of glucose oxidase generally results in release of the flavin). Glucose oxidase is a relatively stable enzyme and can be stored for several years at 0°C in the form of a lyophilized powder.

5.2 Measurement of glucose oxidase activity

Glucose oxidase activity can be determined using a spectrophotometric assay based on a peroxidase indicator reaction that measures the hydrogen peroxide liberated in the enzymatic reaction. Details of a standard assay procedure are given below.

Protocol 6. Determination of glucose oxidase activity

1. Dissolve 0.53 g of 3,5-dichloro-2-hydroxy-benzenesulphonic acid in distilled water and adjust the pH to 7 by addition of 1 M NaOH. Add 30 mg of horseradish peroxidase and make the final volume up to 100 ml (reagent 1). This solution is stable for 4 weeks at 4°C.
2. Dissolve 162 mg of 4-aminophenazone in 100 ml of distilled water (reagent 2).
3. Prepare 0.133 M sodium phosphate buffer containing 7 mM EDTA (reagent 3) by dissolving the following in 1 litre of distilled water: 14 g of disodium hydrogen orthophosphate dihydrate; 8 g of sodium dihydrogen orthophosphate and 0.372 g of EDTA. The pH of the solution should be in the range pH 6.9−7.1. Store as reagent 1.
4. Dissolve 18 g of glucose in 100 ml of distilled water (reagent 4). Store as reagent 1.
5. Prepare a sufficient quantity of a composite assay reagent—for each measurement 1.95 ml is required and has the following composition:
   - 1.55 ml of reagent 3
   - 0.20 ml of reagent 4
   - 0.20 ml of reagent 1
   This composite should be prepared on the day of use.
6. Place 1.95 ml of the combined reagent solution in a 3 ml cuvette and add 0.05 ml of reagent 2. Measure the background absorbance at 520 nm, then add 0.05 ml of an appropriately diluted solution of glucose oxidase and measure the change in absorbance (ΔA) at 520 nm over the initial 2 min.
7. The enzyme activity in units (where 1 unit is defined as the amount of enzyme that causes the oxidation of 1 μmol of glucose per minute under the given conditions) can be calculated as follows:
   Assuming the extinction coefficient of the dye is 13 300 M^{-1} cm^{-1} (*note*: 2 moles of hydrogen peroxide are required to yield 1 mole of dye).

$$[DYE] = \frac{\Delta A}{13.3 \times 10^3 \times 2} \text{ mol l}^{-1} \text{ min}^{-1}$$

$$= \frac{\Delta A \times 10^6}{13.3 \times 10^3 \times 2} \mu\text{mol l}^{-1} \text{ min}^{-1}$$

$$= \frac{\Delta A \times 0.05}{26.6} \mu\text{mol min}^{-1}$$

The basis of this particular assay is the oxidative coupling of hydrogen peroxide with 4-aminophenazone and a phenol, in the presence of horseradish peroxidase to yield a chromogen (a quinoneimine dye) with a maximum absorption at 520 nm:

$$2H_2O_2 + \text{4-aminophenazone} + \text{Phenol} \rightarrow \text{Quinoneimine dye} + 2H_2O$$

Although commercial preparations of glucose oxidase may contain as little as 0.01% catalase as contaminant, the high turnover number for this latter enzyme means that it will be an efficient competitor for hydrogen peroxide in any assay of the type described above. It is therefore essential to determine the catalase activity of a glucose oxidase preparation before it is used.

5.3 Electrochemical detection of glucose oxidase

The ideal method of detecting the presence of a redox enzyme by electrochemical means would be to examine its electrochemistry directly (assuming no other species in the sample were also electroactive in the potential range of interest). However, the direct electrochemistry of enzymes at electrodes is difficult to achieve and therefore small-molecule, electroactive species are used to mediate electron transfer between enzyme and electrode. A variety of non-physiological electron acceptors for glucose oxidase have been studied in this capacity e.g. hexacyanoferrate (III), $N,N,N'N'$-tetramethyl-4-phenylenediamine, benzoquinone, ortho-chloranil and 2,6-dichlorophenolindophenol. These compounds are not used routinely however as they exhibit one or more of the following unfavourable characteristics: rapid autoxidation of the reduced form; cytotoxicity; instability in the reduced form; pH-dependent redox potentials. An alternative class of mediators that does not suffer from these disadvantages includes ferrocene [bis(η^5-cyclopentadienyl)iron Fe(Cp)$_2$] and its derivatives (10).

In order to use an enzyme as a label in an immunoassay it is necessary to be able to detect the catalytic activity of small amounts of the enzyme. DC cyclic voltammetry is a useful diagnostic tool for examining enzyme catalysed reactions. The extent of perturbation of the voltammogram will depend on the rate of the enzyme reaction compared with the time taken to perform the experiment. For a catalytic mechanism of the type:

$$R \rightarrow O + ne^-$$

$$O + X \rightarrow R + Y$$

component X will convert O back to R, at potentials where O is being generated at the electrode, through oxidation of R. The result is an increase in the anodic current and a decrease in the cathodic current. In the absence of label, in this case glucose oxidase, a reversible cyclic voltammogram of the mediator is obtained. However, on addition of enzyme, and assuming an excess of substrate, there is a distinctive change in the form of the voltammogram and catalytic current flows. A number of other examples of enzymes that react with ferrocenes can be found in Chapter 2.

5.4 Preparation of enzyme conjugates

5.4.1 Glucose oxidase-labelled antibody

Glucose oxidase is a glycoprotein containing up to 16% carbohydrate and is therefore ideally suited to direct conjugation using the periodate ($NaIO_4$) method. This procedure is based on the principle that periodate treatment of the polysaccharide moiety of the enzyme will result in the formation of active aldehyde groups which can then react with the amino groups of the antibody to form Schiff bases. These compounds are extremely labile and have to be stabilized by reduction with, for example, sodium borohydride.

A detailed experimental procedure for conjugation of glucose oxidase to IgG is given below.

Protocol 7. Preparation of a glucose oxidase – IgG conjugate using the periodate method

1. Dissolve 16 mg of high purity glucose oxidase in 1 ml of water.
2. Immediately prior to use prepare a solution of 100 mM sodium meta-periodate in water. Add 0.2 ml of this to the enzyme solution and stir for 20 min at room temperature.
3. Transfer the enzyme solution to a dialysis sac and dialyse against 1 mM sodium acetate buffer, pH 4.4, overnight at 4°C.
4. Dissolve 8 mg of IgG in 1 ml of 10 mM carbonate buffer, pH 9.5.
5. Remove the enzyme from the dialysis sac and add to it the solution of IgG. Stir the resulting mixture for 2 h at room temperature.
6. Prepare a fresh solution of 4 mg ml^{-1} sodium borohydride in water.
7. Add 0.1 ml of the sodium borohydride solution to the enzyme – antibody solution and leave stirring gently for 2 h at 4°C.
8. Pass the enzyme – antibody solution down a column of Sephadex G-25 pre-equilibrated with 20 mM sodium phosphate buffer, pH 7. This removes unreacted reagents and adjusts the pH to one that is more favourable for glucose oxidase—the enzyme is not stable for any length of time at basic pH.
9. Store the dilute glucose oxidase – IgG conjugate at 4°C. Freezing of the solution is not recommended.

There are a number of factors that can affect the outcome of the conjugation experiment, including the following:

(a) the concentration of the species to be conjugated. The law of mass action states that the rate of complex formation is proportional to the concentration of the reactants. Thus, the more dilute the reactant solutions, the longer the incubation time required to obtain the same amount of conjugate;

(b) the ratio of the molar concentrations of the enzyme and antibody. The formation of different conjugates follows a Poisson distribution;
(c) the nature of the buffer. The pH and ionic strength should be adjusted so as to maximize the probability of achieving conjugation as desired;
(d) the purity of reagents.

Oxidation of glucose oxidase with periodate has little effect on enzyme activity and heat stability. However, the oxidized enzyme is less resistant to detergents than the native enzyme.

5.4.2 Glucose oxidase – hapten conjugates

The conjugation of glucose oxidase to haptens may be achieved using the mixed anhydride method (see Section 4.5.1). The procedure for preparation of a glucose oxidase – thyroxine(T4) conjugate is described below.

Protocol 8. Preparation of a glucose oxidase – thyroxine conjugate

Preparation of methyl thyroxinate hydrochloride

1. Prepare dry methanol by distilling from magnesium and store over an activated molecular sieve (3 Å). Saturate the dry methanol with HCl gas.
2. Dissolve 1 g of thyroxine in the methanol – HCl and leave the mixture at room temperature overnight. The ester will precipitate during this period.
3. Filter the product; wash with acetone and store in a vacuum desiccator.

Preparation of the succinic anhydride derivative

1. Dissolve 300 mg of methyl thyroxinate hydrochloride and 500 mg of succinic anhydride in 9 ml of tetrahydrofuran (THF)/dimethylsulphoxide containing 0.6 ml of triethylamine. Allow the reaction to proceed for 30 min.
2. Precipitate the product by addition of excess distilled water and collect by filtration.
3. Dissolve the solid material in acetone and then precipitate with hexane.

Preparation of the enzyme conjugate

1. Add 10 mg of the succinic acid derivative to dry THF (cooled to $-5\,°C$), with stirring. (*Note*: it is essential that the THF be dried thoroughly before use).
2. Add 15 μl of triethylamine and 13 μl of isobutylchloroformate. Stir for 30 min, whilst keeping the mixture dry and at $-5\,°C$.
3. Allow the reaction mixture to warm up to room temperature and stir for a further 60 min.
4. Dissolve 110 mg of glucose oxidase in 50 ml of 0.1 M sodium carbonate solution and then add the THF solution dropwise, with stirring.

Protocol 8 continued

5. Stir the resulting solution for 24 h at room temperature, then add 5 ml of 1 M glycine solution and stir for a further hour.
6. Remove all solid material by centrifugation; add FAD to a final concentration of 1 μM and dialyse the solution, firstly against water and then 20 mM Tris−HCl, pH 7.5 containing 0.1 M NaCl.
7. Concentrate the final solution to a suitable volume; filter and purify further by gel filtration on Sephadex G-25.

The first step is conversion of the free acid to methyl thyroxinate hyrochloride, followed by synthesis of the succinic anhydride derivative. This compound can then be used in the mixed anhydride reaction to form an enzyme conjugate.

5.5 Preparation of hapten – mediator conjugates

An advantage of using ferrocene derivatives as mediators is that a wide range of synthetic chemistry may be performed with such compounds. For example, it is possible to configure an electrochemical immunoassay whereby the hapten is labelled with ferrocene and glucose oxidase is added as a separate reagent. Experimental details for the preparation of a theophylline−ferrocene conjugate starting from 5,6-diamino-1,3-dimethyluracil and ferrocene acetic acid are given below.

Protocol 9. Preparation of a ferrocene − theophylline conjugate

1. Dissolve 16.7 g of 5,6-diamino-1,3-dimethyluracil and 20 g of ferrocene acetic acid in N,N'-dimethylaniline and reflux for 5 h using a Dean-Stark water separator.
2. Cool the mixture and add, with stirring, ~400 ml of 8% (w/v) sodium hydroxide solution. Leave the mixture stirring overnight.
3. Remove the N,N'-dimethylaniline by steam distillation. This will leave a yellow precipitate which should partially dissolve on acidification with concentrated hydrochloric acid.
4. Collect the solid material by filtration (the filtrate should be a red solution) and wash the yellow powder with acetonitrile followed by ether.

As discussed in Section 4.5.1, modification of the hapten should not be at the antigenic site—in this particular example the 8-position on theophylline is the point of linkage to the ferrocene moiety. However, it should be noted that this may not be appropriate for use with a different anti-theophylline antiserum.

5.6 Configuration of electrochemical immunoassays

The following sections outline various electrochemical immunoassay configurations in which detection of label is dependent upon the ability of ferrocene derivatives

Immunoelectrodes

to act as mediators to glucose oxidase. There are a number of practical details that should be noted when working with such systems:

(a) glucose solutions should be prepared in buffer and allowed to equilibrate overnight into the α and β anomers. The solutions may be stored for up to a week at 4°C;

(b) stock solutions of ferrocene derivatives should be prepared daily;

(c) many ferrocene derivatives are only sparingly soluble. It is not advisable to attempt to increase the solubility of these compounds by the addition of organic solvents as this will lead to changes in observed electrochemistry. Also, for some enzymes even low concentrations of organic solvent can be highly inhibitory to enzyme activity. Dissolution in a small quantity of a surfactant such as 0.5% (v/v) Tween 20 (Sigma) is an acceptable alternative.

5.6.1 A Sandwich assay using a glucose oxidase – antibody conjugate

The glucose oxidase – antibody conjugate can be used in a conventional sandwich assay format. The detection system can be based on the use of a ferrocene derivative as an electron transfer mediator and glucose as substrate for the enzyme label. For convenience, the assay can involve the use of beads as the solid phase which can then be readily transferred to an electrochemical cell after the binding and washing steps have been performed. A suitable analyte for estimation in this way is human choriogonadotrophin.

5.6.2 A competition assay using a glucose oxidase – T4 conjugate

In this assay free enzyme-modified T4 is measured after competition between modified and unmodified T4 for a fixed number of antibody binding sites. The binding of enzyme-labelled T4 to antibody inhibits the enyzme activity. Thus, in the presence of mediator and glucose, the binding of conjugate to antibody will cause a detectable change in the electrode current. This decrease in current may be used to give a quantitative determination of the analyte.

5.6.3 A competition assay using a ferrocene – theophylline conjugate

Drug – ferrocene conjugates will act as effective electron acceptors for glucose oxidase. However, the catalytic current produced in the enzymatic oxidation of glucose is inhibited upon binding the drug – ferrocene complex with antibody to the drug (*Figures 9* and *10*). The binding of low molecular weight, rapidly diffusing species to an immunoglobulin results in a large molecular weight complex that is no longer rapidly diffusing and therefore cannot act as a mediator to glucose oxidase. An assay may be based upon the reversal of this effect on addition of non-labelled drug. *Figure 11* shows a calibration for theophylline.

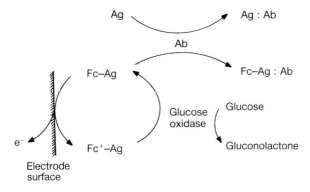

Figure 9. Schematic representation of the ferrocene-mediated electrochemical immunoassay. Ag is antigen (i.e. drug); Fc – Ag is the ferrocene – drug conjugate and Ab is antibody to the drug.

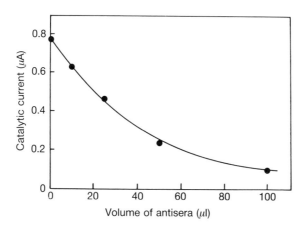

Figure 10. Inhibition of the catalytic current on addition of varying volumes of anti-theophylline antiserum to conjugate/glucose oxidase/glucose mixtures.

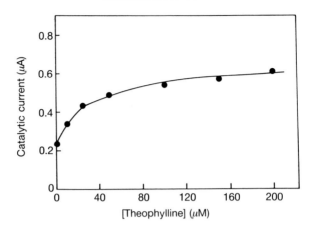

Figure 11. Calibration for theophylline using the ferrocene-mediated electrochemical immunoassay.

References

1. Janata, J. and Blackburn, G. F. (1984). *Ann. N.Y. Acad. Sci.*, **428**, 286.
2. Frew, J. E., Green, M. J., Hopkins, A. R., and Mannin, C. J. (1988). In *Complementary Immunoassays*. Collins, W. P. (ed.), John Wiley, Chichester, p. 209.
3. The Southampton Electrochemistry Group (1985). *Instrumental Methods in Electrochemistry*. Ellis Horwood, Chichester.
4. McComb, R. B., Bowers, Jr., G. N., and Posen, S. (1979). *Alkaline Phosphatase*. Plenum Press, New York.
5. Bard, A. J. and Faulkner, L. R. (1980). *Electrochemical Methods. Fundamentals and Applications*. John Wiley, New York.
6. Heineman, W. R. and Halsall, H. B. (1985). *Anal. Chem.*, **57**, 1321A.
7. Tijssen, P. (1985). *Practice and Theory of Enzyme Immunoassays*. Burdon, R. H. and van Knippenberg, P. H. (ed.), Elsevier, Amsterdam.
8. Kemeny, D. M. and Challacombe, S. J. (eds) (1988). *ELISA and Other Solid Phase Immunoassays: Theoretical and Practical Aspects*. John Wiley, Chichester.
9. Dean, P. D. G., Johnson, W. S., and Middle, F. A. (eds) (1985). *Affinity Chromatography: A Practical Approach*. IRL Press, Oxford.
10. Cass, A. E. G., Davis, G., Francis, G. D., Hill, H. A. O., Aston, W. J., Higgins, I. J., Plotkin, E. V., Scott, L. D. L., and Turner, A. P. F. (1984). *Anal. Chem.*, **56**, 667.

5

Conductimetric and impedimetric devices

DOUGLAS B. KELL and CHRISTOPHER L. DAVEY

1. Introduction

The idea that (changes in) the electrical conductivity or impedance of a biological (or other) system may be used to sense biological activity goes back to the previous century (1). Notwithstanding, and despite many spectacular successes which include the discovery of the molecular thickness of biological membranes and the voltage-gated conductivity of nerve axons, the generalized impedimetric approach ('impedance or admittance spectroscopy') remains astonishingly underexploited. Thus, Campbell and Dwek's eponymous and otherwise excellent book, *Biological Spectroscopy* (2), does not even mention the technique. In the belief that this widespread lacuna is due to a general lack of appreciation of the relevant theory and terminology, we begin this chapter by a consideration of exactly what it is that is measured in an impedimetric experiment, and introduce the multifarious terminologies that many workers have applied to essentially the same phenomena. Given a proper appreciation of this, the practical approaches actually used, together with their pitfalls, follow naturally.

It must be admitted, at the time of writing, that the exploitation of impedimetry in biosensors in their modern sense is in its very infancy. Nonetheless, we take the view that provided an adequate grasp of the theory and experimental background is to hand, the development of biosensing devices for desired goals also follows naturally. To this end, we also illustrate the potential of the technique with some theoretical and experimental devices.

2. Theory of the impedimetric experiment

2.1 The concepts of impedance and admittance

The generalized impedimetric experiment may be described with reference to *Figure 1*. In this, a small-amplitude, sinusoidally modulated voltage $V_m \sin \omega t$, of frequency ω rad sec^{-1} (= $2\pi f$ Hz), is applied to the system of interest. In the (quasi-)steady state, after any transients have decayed, a current then flows in the system with a time-dependence $i_m \sin(\omega t + \theta)$, where θ is the so-called phase angle.

Conductimetric and impedimetric devices

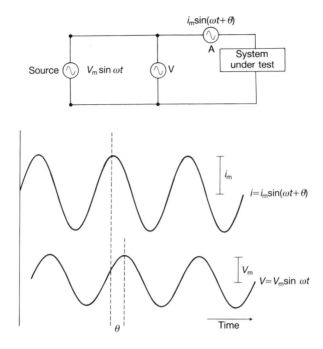

Figure 1. The generalized impedimetric experiment, in which a sinusoidally-modulated waveform is applied to a system under test. The resultant current flow is of the same frequency but shifted in phase. As described in the text, the (linear) impedance is obtained from the measured current, voltage and phase angle. The waveforms and their phase relationship are shown.

Thus, at least if the system is linear in the sense that the current flowing is proportional to the voltage applied, the current and the voltage are sinusoids of the same frequency, and the impedance **Z** is uniquely determined by the macroscopic observables V_m, i_m, and θ.

So far as the experimenter is concerned, and except in unusual cases (outside the scope of this article) in which inductive elements are present, the system under test may be taken to behave (at a given frequency) either as a resistor and capacitor in series or as a conductor and capacitor in parallel. In the former case, we measure the impedance of the system as the dependence of the voltage on the current, and treat the system as being connected to a current source of infinite resistance. In the latter case, the system under test is taken to be connected to a voltage source of zero resistance and the admittance describes the dependence of the current upon the voltage. This clear separation between the experimenter and the system under test is of the greatest importance, since it is of course the properties of the latter which are of interest.

What is the relationship between the phase angle and the magnitudes of the resistance and capacitance in the system? Physically, one way to consider this is

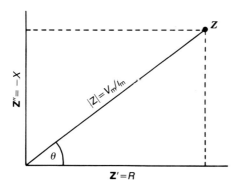

Figure 2. Impedance is a complex quantity, containing both real and imaginary parts. The modulus of the impedance and the phase angle are used to obtain the resistance and reactance according to the relations $R = |Z| \cos\theta$ and $X = -|Z| \sin\theta$.

to realize that all of the energy in the exciting electrical field must either be stored (as a capacitive term) or dissipated (as a resistive term), and that storage takes time whilst dissipation to heat is instantaneous on the time-scale of present interest. Thus for a pure resistor the phase angle is zero, whereas if the system under test is a pure capacitor the current leads the voltage by $\pi/2$ rad, i.e. the voltage is 90° out-of-phase with the current. For real systems, which at a given frequency behave as a resistor and capacitor in series (or parallel), the phase angle takes a value between zero and 90°. Now, impedances and admittances are complex quantities (in the sense that they contain both real and imaginary parts), the impedance, which is a vector quantity, being given by $Z = R + jX$ where R is the resistance, X the reactance ($= -1/\omega C$) and $j = \sqrt{-1}$. The relationship between Z, V_m, i_m and θ is given in *Figure 2*. $|Z|$ is known as the modulus of the impedance and θ the argument. From simple trigonometrical considerations it may be observed that $R = |Z| \cos\theta$, $X = -|Z| \sin\theta$ and $Z^2 = R^2 + X^2$. The units of Z, R and X are ohms (Ω) whilst the SI unit for capacitance is the Farad (F). The admittance Y is the reciprocal of the impedance (i.e. $Y = 1/Z = G + jB$ where G is the conductance and B ($= \omega C$) the susceptance of the system. The units of Y, G and B are Siemens (S). Since these concepts of impedance, admittance, conductance, resistance, susceptance and reactance are only different ways of treating the system under test, it is obvious that they are related to each other in some way, and *Table 1* groups together the relevant relationships. It may be noted, for instance, that resistance = 1/conductance only when the reactance or susceptance = zero and not simply when the capacitive terms are ignored by splitting the impedance/admittance into its real (in-phase) and imaginary (90° out-of-phase) components.

As mentioned above, any linear system may be treated, at a given frequency, as consisting of a resistor and capacitor in series or in parallel, regardless of the actual

Table 1. The relationships between impedance and admittance, and their real and imaginary parts.

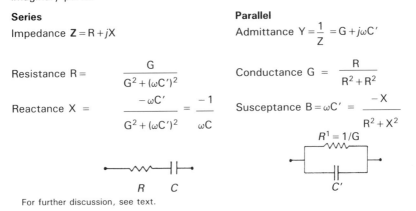

For further discussion, see text.

complexity of the equivalent electrical circuit. Distinguishing the appropriate equivalent electrical circuit is only possible if the frequency is varied, and it is because one studies the absorption of electrical energy (electromagnetic radiation) by the system as a function of frequency that the generalized impedance technique is often referred to as Impedance Spectroscopy (3), Admittance Spectroscopy (4) or Dielectric Spectroscopy (5). Plots of the imaginary part of the impedance or admittance versus the real part (with frequency as the parameter) tend to give semicircles, each semicircle corresponding to a capacitive element in the admittance domain and with the time constant τ given by RC (which has units of seconds). In some cases, several different circuits may be used to fit the same data and physical intuition based on knowledge of the physical structure or organization of the system is required to determine the appropriate equivalent electrical circuit.

2.2 Intrinsic system properties and dielectric relaxation

If our system consists of a block of condensed matter connected to the measuring device via two plane-parallel electrodes, the currents measured in response to an exciting voltage will depend upon the electrode area, A, and the separation between them, d. To normalize the intrinsic properties of the system between the electrodes, we define a conductivity σ' and permittivity ϵ', which are related respectively to the conductance G and capacitance C by:

$$\sigma' = G\,(d/A) \qquad (1)$$

$$\epsilon' = C\,(d/A\epsilon_0) \qquad (2)$$

where ϵ_0 is the capacitance of a unit cell containing a vacuum, usually known as the 'permittivity of free space' and with a numerical value equal to 8.854×10^{-12} F m^{-1}. σ' has units of S m^{-1} whilst ϵ' is dimensionless. The factor (d/A) has the dimension length^{-1} and is known as the cell constant. Water at 37°C has a permittivity of some 74.4, so it may be calculated from equation 2 that a unit cell containing it has a capacitance of some 6.59 pF.

As with impedance and admittance, one may define a complex permittivity $\epsilon^* = \epsilon' - j\epsilon''$ which has both real (ϵ') and imaginary (ϵ'') parts. ϵ'' is known as the dieletric loss and is related to the conductivity by:

$$\epsilon = (\sigma' - \sigma'_L)/\omega\epsilon_0 \qquad (3)$$

where σ'_L is the DC or 'low-frequency' conductivity of the system which, in the case of biological systems will tend to be the conductivity of (i.e. due to) the small inorganic ions present.

The passive electrical behaviour of biological systems generally changes, as a function of the frequency, in between 'plateau' values ϵ'_L and ϵ'_∞ (although there may be significant overlap), each region of frequency-dependence being known as a 'dispersion'. A substance whose passive electrical behaviour is that of a parallel RC circuit having but a single relaxation time obeys the Debye equation:

$$\epsilon^* = \epsilon'_\infty + \Delta\epsilon'/(1 + j\omega\tau) \qquad (4)$$

$$\text{where } \Delta\epsilon' = \epsilon'_L - \epsilon'_\infty, \qquad (5)$$

$$\epsilon' = \epsilon'_\infty + \Delta\epsilon'/(1 + \omega^2\tau^2) \qquad (6)$$

and

$$\epsilon'' = \Delta\epsilon'\omega\tau/(1 + \omega^2\tau^2) \qquad (7)$$

Inspection of these equations reveals that ϵ' is a monotonically decreasing function of frequency (between two 'plateau' regions), ϵ'' exhibits a peak at the frequency at which $\epsilon' = \epsilon_\infty + \Delta\epsilon'/2$, and that a plot of ϵ'' versus ϵ' gives a circle of radius $\Delta\epsilon'/2$ and whose centre is located on the abscissa, with the maximum value of ϵ'' occurring when $\omega\tau = 1$. This is illustrated in *Figure 3*.

In practice, real systems give semicircles whose centre lies below the abscissa, according to an empirical modification of the Debye equation due to the Cole brothers (6):

$$\epsilon^* = \epsilon'_\infty + \Delta\epsilon'/(1 + j\omega\tau)^{(1-\alpha)} \qquad (8)$$

Equation 8 is widely used in biological impedance work, with α usually being taken to reflect some type of distribution of relaxation times. Certainly the fit of data to equation 8 is generally very good, but as pointed out more than 30 years ago by Schwan (7) a whole host of relaxation-time distributions give behaviour virtually indistinguishable from each other. In other words, the use of α as a free variable permits most data to be fitted, regardless of the physical mechanisms at

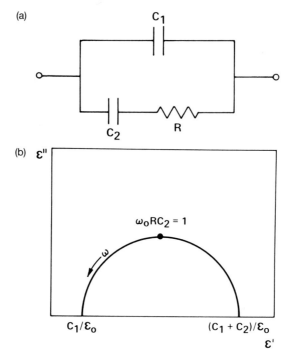

Figure 3. (a) An equivalent electrical circuit for a substance having but a single relaxation time, such as a system exhibiting Debye behaviour. (b) A complex permittivity plot of the dielectric data obtained therefrom.

work. Jonscher (8) and Dissado and Hill (9) in particular have argued that the wide 'distribution of relaxation times' commonly observed in work with solid-state systems is better ascribed to a hierarchical interaction of the relaxing particles with the matrix in which they are embedded, a view with which we have much sympathy since at least in enzymology it is common to assume that proteins made transitions between different conformational macrostates sequentially and not in parallel (10–12).

What are the molecular mechanisms of a dielectric dispersion? The simplest is that of the rotation of a molecule with a permanent dipole moment as shown in *Figure 4*. Here we illustrate a single molecule, which we will take to be globular protein, with a permanent dipole moment along the direction of the arrow, representative of an ensemble of the same and held between two electrodes connected to an AC source. If this dipole consists of single elementary charges (q^+, q^-) of opposite sign separated by a distance s, the molecular dipole moment m, which is an intensive property, is given by $m = qs$, with units of C m. (For historical reasons, dipole moments are often given in the non-SI unit Debyes (D), where $1D = 3.33 \times 10^{-30}$ C m, and the displacement of a single electronic charge through 10^{-10} m

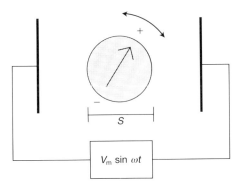

Figure 4. Rotation of a molecule with a permanent dipole moment as a mechanism of dielectric relaxation.

gives a dipole moment of 4.8 D.) Proteins typically have permanent dipole moments of a few hundred D, equal to some 3–5 elementary charges separated by the molecular diameter (13).

If we apply a sinusoidally modulated electrical field to the protein, it will attempt to rotate so that the negative charge is facing the positive electrode and vice versa. Because the net charge on the protein is zero, no DC electrophoresis is possible. If the frequency of the field is low, the protein will have time to orient, and this will be accompanied by a displacement current. When the direction of the field is reversed the protein will reorient to face the opposite electrode. In each case, the storage of charge by the oriented protein may be represented as a leaky capacitor. If the frequency of the electrical field is high enough, however, the protein will not have time to reorient, since it takes time to overcome the viscous drag exerted by the solvent bath. The rotational relaxation time for a (hard) sphere is given by:

$$\tau = \xi/2kT \tag{9}$$

where ξ is a molecular friction coefficient, k is Boltzmann's constant and T the absolute temperature.

If we take the protein to be equivalent to a hard sphere of radius a, turning in a Newtonian fluid of viscosity η, the Stokes–Einstein relation holds:

$$\xi = 8\pi\eta a^3$$

so that

$$\tau = 4\pi\eta a^3/kT \tag{10}$$

Thus we may expect (from equations 5 and 6), and indeed find, that the capacitance (or permittivity) of a protein solution is frequency-dependent, with the form of an inverted sinusoid when plotted against the logarithm of the frequency, and with the

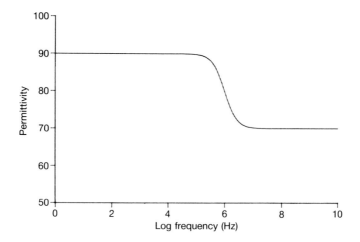

Figure 5. A simulated dielectric spectrum of a hard spherical protein containing a permanent dipole moment. For the simulation, $\Delta\epsilon'$ was taken to be 20 permittivity units for a protein concentration of 100 mg ml^{-1}, and f_c to be 1 MHz. The dispersion follows equation 5.

dispersion being half-complete at a 'characteristic' frequency $f_c = (2\pi\tau)^{-1}$. For fundamental reasons (often referred to as the Kronig–Kramers relationships), the conductivity must increase as the permittivity decreases (since the energy in the field must either be stored or dissipated) and the dielectric increment $\Delta\epsilon'$, the conductivity increment $\Delta\sigma'$ and the relaxation time τ for a single dielectric dispersion are related by:

$$\tau = \Delta\epsilon'\epsilon_0/\Delta\sigma' \tag{11}$$

Equation 10 tells us the relaxation time but not the dielectric increment. This latter quantity depends upon the molecular dipole moment m, the concentration of the protein, and other factors according to:

$$\Delta\epsilon' = Ncgm^2/2\epsilon_0 MkT \tag{12}$$

where N is Avogadro's number, c the protein concentration in mg ml^{-1}, M the protein's molecular weight and g a parameter introduced by Kirkwood (14) to account for molecular associations.

A simulated dielectric spectrum of a protein is given in *Figure 5*, where it may be observed that the dielectric increment is some 20 permittivity units per 100 mg ml^{-1} of protein, with a characteristic frequency of about 1 MHz, values which fit those to be expected on the basis of equations 10 and 12. Non-spherical globular proteins gives dielectric spectra whose Cole/Cole $\alpha > 0$; in cases of a very high axial ratio, separate relaxations of the two ellipsoids of revolution may be discerned (15). From the biosensing point of view, it should be fairly evident (since $\Delta\epsilon'$ is proportional to concentration) that simple permittivity measurements of protein

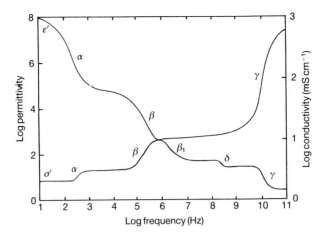

Figure 6. Classical dielectric dispersions exhibited by a typical biological tissue.

solutions allow the measurement of their concentration, a fact of obvious utility in all sorts of process streams (16). If a protein of known properties is present behind a dialysis membrane, a novel and non-invasive dielectric probe of microviscosity may also be imagined, since the measurable relaxation time for protein rotation will depend (inversely) upon the local (micro)viscosity.

Although the above theory represents the briefest that can be sensibly given, and readers are urged to consult the fuller treatments presented elsewhere (3-5,4,7,8,17-19), it will suffice for present purposes. Before we go on to the technical aspects, however, it is worth considering the mechanisms of dielectric relaxation exhibited by biological systems at all levels of organization.

2.3 Mechanisms of dielectric relaxation in biological systems

A 'classical' dielectric spectrum of a biological tissue is illustrated in *Figure 6*. This shows three major and two minor dispersions, mainly attributed to: tangential flow of ions along cell surfaces (α), build-up of charge at cell membranes via a Maxwell-Wagner effect (β), rotation of small molecular weight dipoles, especially water (γ), rotation of macromolecular side-chains and 'bound' water (δ), and protein rotation (β_1). Not shown is a lower-frequency μ-dispersion (20-22) observable in some membrane vesicle preparations and ascribed to various field-induced motions of membrane lipids and proteins. The dielectric behaviour of mixtures or of emulsions differs depending upon whether they are oil-in-water or water-in-oil emulsions (23,24), and we give an example of the former in Section 4.2. The important points are

(a) leaving aside electrode effects which are dealt with later, *any* permanent or induced dipole may give rise to a dielectric dispersion;

Conductimetric and impedimetric devices

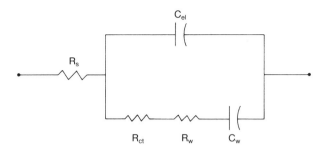

Figure 7. The Randles equivalent circuit for electrochemical systems. C_{dl} is a double-layer capacitance, R_{ct} represents the charge-transfer resistance (due to the Faradaic reaction) and R_w and C_w together represent the 'Warburg' impedance which accounts for the difficulty of bringing electroactive substances to and from the reaction layer. R_s represents the bulk, series resistance between the electrodes. The structures represented by the capacitive elements are heterogeneous, and this is reflected by the symbols used.

(b) if the dipole's magnitude or relaxation time changes as a function of time or of the concentration of an added ligand or determinand then a sensing principle is available;

(c) the frequency range available covers many orders of magnitude, and indeed far more than that of any other spectroscopic technique.

Except for very-high-frequency work, which we cover but little here, electrodes are necessary to connect the measuring device to the sample of interest. This is because the current carriers in the wires are electrons but, since the hydrated electron has a very short lifetime in solution, it is necessary that ions carry the current in the aqueous media characteristic of biological systems. The electrodes thus represent (and are defined as) the interfaces between the measuring system and the measured system. However, the measuring system does not know this, so that what we measure in a system such as that in *Figure 1* is the behaviour of the electrodes *plus* the biosystem. Whilst this may work either to our advantage or to our disadvantage in biosensing devices, it is at least necessary to have some idea of the important topic of electrode choice and preparation.

3. Electrodes

3.1 Polarizable and non-polarizable electrodes

Electrodes are generally classified into polarizable and non-polarizable. Of course this qualitative distinction admits a quantitative description, and there is no genuinely non-polarizable electrode, only an approximation to it. Polarization describes the fact that when a current is forced to cross an electrode–electrolyte interface there will be a resistance to such current flow and thus a tendency for a potential drop to occur across the interface. If the current is alternating, there can also be a phase

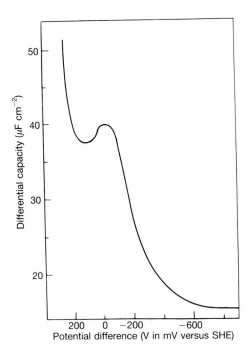

Figure 8. Potential dependence of the differential capacitance of a mercury electrode in 0.1 M HCl. Note the large values observed in certain regions of potential.

shift between the current and the voltage. Some of the many processes that can contribute to this potential drop are embodied in the 'Randles' equivalent circuit for an electrode–electrolyte interface shown in *Figure 7*. The chief of these is the double-layer capacitance (C_{dl}), which is set up by virtue of the fact that there must be a molecular discontinuity at the phase boundary between the electrode and the 'bulk' solution, such that there will be a tendency for ions of a charge opposite to that of the electrode to populate the compact and diffuse layers adjacent to the electrode surface. The electrode surface and the solution side of the double layer represent two plates of a capacitor, and since C is inversely related to the distance between a capacitor's plates the double layer capacitance can be very large, typically some 10 $\mu F\ cm^{-2}$. The absolute charge on an electrode surface depends upon the electrode material, *inter alia*, so that it is a function of the mean electrode potential. *Figure 8* shows a typical plot of the so-called differential capacitance (at ~1 kHz) versus potential for Hg in 0.1 M HCl. Factors contributing to the exact magnitude of C_{dl} in solid electrodes are not well-understood to this day.

As well as the bulk solution resistance (R_s), *Figure 7* also shows a charge-transfer resistance (R_{ct}) and a 'Warburg impedance' (Z_w, composed of a Warburg resistance and reactance). R_{ct} represents the resistance to current flow due to the difficulty of carrying out the Faradaic electrochemical reaction itself, whilst Z_w represents the semi-infinite linear diffusion of electroactive substance up to the reaction layer. Typical DC amperometric measurements, in which a potential is set (at a value such that R_{ct} is negligible) and a current is measured so that their ratio effectively constitutes a DC resistance, are dominated by Z_w. In the absence of impurities, a constant current source will adjust the relevant potentials to permit the attainment of the desired current by electrolysis. Thus polarizable electrodes are those (such as Hg) which have a high overvoltage for current flow due to H_2 evolution (or, if this is more limiting, for O_2 evolution). Non-polarizable electrodes are those such as Ag/AgCl/Cl$^-$ in which the ability of AgCl to act as both an electronic conductor and a participant in the Faradaic reaction $AgCl + e^- \rightarrow Ag + Cl^-$ is cleverly exploited. We next consider the impedance of elemental, and particularly of metallic electrodes.

3.2 The impedance of elemental electrodes

From the discussion above, it is clear that to minimize the impedance due to electrode polarization, large-area electrodes with a reversible electrochemistry and a clean surface are required. The area of the electrode that is of interest here is the actual surface area, and not the geometric area. In general, the noble metals gold, platinum and palladium exhibit the greatest degree of reversibility of electrolytic reactions, although the quantum theory of electrochemistry, and the constituent processes

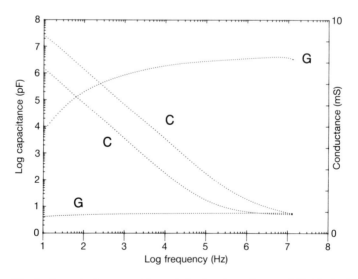

Figure 9. Effect of conductance on the polarization impedance of Pt black electrodes. Measurements were made using a Hewlett-Packard Model 4192A Impedance Analyser (41), in 10 mM and 100 mM solutions of KCl. The polarization capacitance (increase) and conductance (drop) at low frequencies are the more marked for the more conductive solution.

involved in even such a 'simple' case as water electrolysis, are not understood to a level sufficient to permit a calculation of the 'best' metal from first principles. Metals such as titanium, aluminium and iron, which tend to form non-conducting oxide layers on their surfaces, exhibit a high impedance, typically increasing substantially below 1−10 kHz. As mentioned above, the impedance is a function of the absolute potential on the metal(s), although in a two-terminal system this is allowed to 'float'. It is also a function of the current density, as illustrated in *Figure 9*, in which the frequency-dependent admittance of two ionic solutions is plotted, using the same Pt black electrodes. Pt black electrodes are Pt electrodes (or other, but Pt forms the best substrate from a mechanical point of view) on the surface of which a colloidal coating of Pt is electrochemically deposited from a solution of platinic chloride. Interference between light rays reflected from the very grainy surface of this electrode causes it to appear black. Because Pt black electrodes have a special place in biological impedimetry, we consider them in some detail.

3.2.1 Preparation of Pt black electrodes
Despite more recent studies, the general method described by Kohlrausch in the 1890s is still the best and is described below.

Protocol 1. Preparation of Pt black electrodes

1. Wash the electrode successively with copious quantities of acetone (to remove grease), water and either 1 M hydrochloric or nitric acid.
2. Wash the electrode in doubly distilled, or similarly high purity, water and dry in air.
3. Prepare a plating solution containing 25 mM hydrochloric acid in which are dissolved 3% (w/v) platinum (IV) chloride[a] and 0.025% (w/v) lead acetate.[b]
4. Immerse the clean working electrode into the plating solution along with a large-area platinum foil counter electrode and a Ag/AgCl reference electrode.
5. Attach the electrodes to the appropriate connections on a potentiostat (PAR Model 174A).[c]
6. Adjust the potential applied to the working electrode to give a current density of 10 mA cm^{-2} (typical potentials are in the range 0 to −0.1 V relative to the Ag/AgCl reference electrode).
7. Continue to pass current until a suitable deposit of platinum black has formed. The deposition can be assessed visually and by measuring the impedance − frequency response in a standard ionic solution. Typically a total charge density of 10 C cm^{-2} or less gives a coverage that represents a suitable balance between a low impedance and a mechanically stable coat. However larger charge densities have been used (25).

Protocol 1 continued

^{a} It is important to use platinum (IV) chloride and not the more common potassium chloroplatinate. The latter does not work in this procedure.

^{b} The lead acetate gives rise to a stronger deposit. A mixture of platinum (IV) chloride and lead acetate is known as Kohlrausch's solution and is commercially available.

^{c} Home-built potentiostats can also be used and details on their construction are described in Chapter 3. Other sources of commercial instruments are also given in the Appendix.

To obtain a uniform coating of Pt black, there should in principle be some uniformity in the shape of the field lines joining the counter electrode and the surface of the electrode to be plated. In practice it seems to make little difference, and the sides of pairs of pin electrodes facing away from the counter electrode receive a good deposit.

There is some disagreement about the optimum current densities and charge to be passed for producing a Pt black electrode with the lowest impedance. Schwan (25) gives data to indicate that 30 C cm^{-2} geometric area passed at a current density of 10 mA cm^{-2} provides the best coating. This amount of charge passed represents a very heavy coating indeed, which is mechanically rather weak. A suitable balance between a low impedance and the necessary rate of replating is probably less than one-third of this. The appropriate current density is obtained in the potentiostatic set-up by increasing the cathodic voltage, in our hands typically to a value of approximately 0 to -0.1 V versus Ag/AgCl (3 M KCl). The deposition of Pt is assessed visually, and functionally by checking the impedance–frequency diagram of the electrodes in a standard ionic solution as a function of deposition time.

3.2.2 Other low-impedance electrodes, based on Ag, C and Ir

It is possible to plate Pt black on to other noble metal surfaces (though the adhesion to base metals is generally poor), the method, and surface properties being similar to those for Pt black on platinum. Cole and Kishimoto (26) pointed out that the absolute potential of Pt is rather poorly defined, and sought to construct an electrode that would combine the stable potential of Ag/AgCl with the low impedance of Pt black. To this end they chlorided a silver electrode, using the Ag as the anode in an electrochemical cell containing Cl$^-$ and passing a charge density of 10 C cm^{-2} at 2 mA cm^{-2}. The chlorided Ag electrode was then platinized as in Section 3.2.1 with 1–5 C cm^{-2} at the same current density, and then lightly chlorided again. The platinized AgCl electrode so produced had an equilibrium potential close to that of the original chlorided silver electrode and with an appropriately low impedance. Surprisingly, this electrode has subsequently been little used.

Marmont (27) chlorided a pair of Ag electrodes of a charge density of 2.7–7.5 C cm^{-2} and, reasoning (correctly) that the surface area of the colloidal AgCl would be much greater than that of the Ag base, converted the AgCl to molecular Ag using photographic developer. The impedance (above 300 Hz) of the electrodes so produced was indeed similar to that of the ionic solution between them. This effect is probably due both to an increase in effective electrode area and to the fact that the conductivity of Ag is much greater than that of AgCl (28). Grubbs and Worley (29) observed that the lowest impedance could be obtained by etching the Ag electrode with aqua regia for 30 sec, chloriding at a current density of 50 mA cm^{-2} with a heavy deposit amounting to 6 C cm^{-2}, and dechloriding by reversing the current for an empirically-determined 30 sec such that approximately 25% of the chloride layer was removed. The exact mechanism of this effect is not known.

Although graphite is a good electrical conductor, and glassy carbon is thought by many to provide the most suitable anode for general electrochemical work, it provides an electrode−electrolyte interface which can in fact present an extremely high impedance to the flow of AC current of low (< 10 kHz) frequencies (4). It is not therefore recommended for impedimetric work.

Gielen and Bergveld (30) devised the AIROF (Anodic Iridium Oxide Film Electrode) as a low-impedance electrode, which in the potentiometric mode may be used to sense pH in an O_2-independent fashion, due to its surface chemistry. Ir/O-containing surface films can have greatly different surface behaviours depending on their stoichiometries. These in turn depend upon the method of preparation. The AIROF, which may be used in electrochromic displays since the pertinent redox states of the Ir(OH)$_n$ film are colourless and blue−black (31), is prepared as described below.

Protocol 2. Preparation of an anodic iridium oxide film electrode

1. Prepare a solution of 0.9% (w/v) sodium chloride in water as the electrolyte.
2. Immerse an iridium wire and a large platinum counter electrode into the saline solution and apply a sawtooth voltage with an amplitude of 1.5 V (peak to peak) at a frequency of 10 Hz for 1 min. During this time gas bubbles should appear at the surface of the iridium wire.
3. Immerse the etched iridium wire along with a platinum ring counter electrode and a saturated calomel electrode reference electrode in a solution of 0.5 M sulphuric acid.
4. Cycle the potential applied to the iridium electrode between −1.25 V and +0.25 V with respect to the reference electrode until the increase in current (decrease in impedance) that is observed reaches a steady value. Typically 200−1000 cycles (depending upon the size of the electrode) are necessary. Electrodes prepared in this fashion typically have half the impedance of platinum.

Since it is not possible to be comprehensive here with respect to all types of electrode, we assume that most readers will wish to prepare electrodes of minimum impedance. However, we appreciate that there are cases in biosensing in which one wishes to maximize the electrode impedance whence to derive a ligand-specific signal. In this case one should evidently choose electrodes of small area, covered with a poorly-conducting (e.g. oxide) film and/or the protein/biomolecule of interest. Adsorption of the ligand to the electrode surface will give a significant increase in impedance, and particularly in capacitance, which may be monitored. This approach is discussed in detail by Arwin (32) and others (33,34).

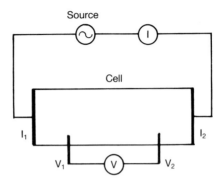

Figure 10. The principle of using four-terminal measurements for the determination of bulk impedances in lossy media. The current electrodes I_1 and I_2 are in series with a pair of voltage electrodes (V_1 and V_2), the latter being connected to a voltmeter with an input of high impedance. This feature means that negligible current flows across the interface of the voltage electrodes so that they cannot polarize.

We have thus far merely stated that the means by which to measure the impedance, conductance or cognate property of a system is to connect it to a sinsuoidally-modulated voltage/current source and measure the current, voltage and phase angle. Whilst this is true, it is not all-encompassing, and before we cover some of the relevant considerations and hardware requirements, we mention an approach which can, in principle, remove the interference of electrode impedances completely.

3.3 Four-terminal systems

In our impedimetric system in *Figure 1*, the electrodes measuring the voltage were the same as those passing the current, that is, we considered a two-terminal system. Since the cause of electrode polarization is the very passage of current across the electrode–electrode interfaces, one way to avoid many of the problems with electrodes is to use two pairs of electrodes (*Figure 10*). The outer pair passes current through the system whilst the inner pair, which are connected across the terminals of a voltmeter of high input impedance, measure the voltage drop across the ('homogeneous') system. If the voltmeter is of high input impedance, negligible current passes the electrode–electrolyte interface of the voltage electrodes so that they cannot polarize. The impedance (of the bulk system) is still the ratio of the vector voltage to the vector current (35), and is determined by the geometry of the inner (voltage) electrodes (36).

4. Hardware requirements

From the hardware point of view, there are many approaches to the measurement of conductance and admittance; early ones are summarized by Schwan (25), and a useful and up-to-date overview appears in the book edited by Macdonald (3). The

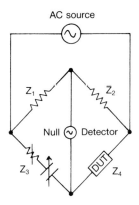

Figure 11. The principle of a two-terminal impedance bridge, in which the device under test (DUT) forms an arm (Z_4) adjacent to that (Z_3) in which a variable impedance is held. At balance, $Z_1 Z_4 = Z_2 Z_3$, and the AC-sensitive null detector registers no current flow.

approaches may best be classified into (i) analogue versus digital and (ii) time-domain versus frequency-domain. The main advantages and disadvantages of each approach relate predominantly to the desired frequency range and to the cost and availability of suitable computing equipment, but at the heart of each lies the aim of measuring or calculating the voltage, the current and the phase angle in a circuit such as that of *Figure 1*, and thus of obtaining the real and imaginary parts of the admittance as described above. Since a discussion of electronic circuit design is outside the scope of this chapter, what follows in this section is merely an overview of the general types of instrument that one might purchase or construct for (relatively low-frequency) impedimetric purposes, together with some comments as to the properties, suitability and limitations of each.

4.1 Frequency-domain methods; analogue

Bridges based on the famous Wheatstone bridge, but including provision for balancing both resistive and capacitive elements, constitute the classical means by which impedances are measured (*Figure 11*). Those devised by Wien, Berberian and Cole, and Calvert (3,25) are the most popular. Their only real use in the modern era is in precision work of the metrological variety (37). In most cases, the limitations on frequency are caused by effects outside the hardware itself, namely instability in the measured system itself (which determine the lowest frequencies at which operation is suitable) and inductive effects within the leads and the electrodes (which depend on the conductance, the square of the frequency and the length of any leads between the sample and the measuring system, and thus determine the upper operating frequency). Independent DC potentiostatic control, for holding the mean potential on the working electrode at a desired value rather than letting it float, is also available with this type of device.

As with all impedimetric methods of the type considered, the linearity assumption is made. In other words, the measured impedance is taken to be independent of the exciting voltage, and this should always be checked. The most suitable AC voltage to use, regardless of any DC potential maintained, is approximately the greatest at which the linearity assumption does hold. Our own experience is that the impedance measured (by looking at the same frequency for both current and voltage) using Pt black electrodes in biological systems is linear by this criterion up to an exciting voltage (peak to peak) of at least 300 mV, though it may be noted that this criterion is less than robust if harmonics are also considered (12,38,39).

Phase-sensitive detectors provide an accurate means of registering the real and imaginary parts of a voltage with respect to a reference voltage. The general mechanism of operation is to multiply the input waveform by a reference (usually square) wave and time-average the output, the result being a voltage which is both phase- and frequency-selective, and which provides an excellent resolution of the small phase angles typically encountered in biological work.

4.2 Digital frequency- and time-domain methods

In the modern era, such methods are probably the methods of choice, since (i) powerful commercial instruments that exploit this approach are available, and (ii) frequency-domain methods have an inherently higher signal:noise method than do time-domain methods (although this advantage is partially offset by the longer measuring times). Although each type of instrument shares certain properties (and indeed combined instruments such as the Hewlett-Packard model 4195A are becoming available), it has become conventional to class them as Frequency Response Analysers, Impedance (or Network) Analysers and Spectrum Analysers. All exploit digital signal analysis (and in some cases generation) of the appropriate waveforms flowing in the circuit of interest.

Frequency Response Analysers (FRAs), such as the models 1170 and 1250 series produced by Solartron (Schlumberger), are essentially electronic correlators in which the cell response is correlated with two synchronous reference signals, one of which is in phase with the sinusoidal perturbation applied to the system and one of which is shifted by $\pi/2$ rad (40). Under computer control, they provide a convenient approach to the measurement of electrical and bioelectrochemical impedances, and present models cover some 12 decades of frequency (10^{-5} Hz to 10^7 Hz). Present devices, including an electrochemical interface for potentiostatic control and for three- and four-terminal work, cost some £13 000 (1988 prices, excluding computer and software). As with most frequency-domain methods, the sinusoidal waveform is applied one frequency at a time to the system under test, so that, according to the sampling theorem, the minimum measurement time is one half the reciprocal of the lowest frequency of interest.

Automated Impedance (Network) Analysers, such as the Hewlett-Packard 4192, are essentially bridges in which the balancing is carried out automatically. Typically they possess exceptional bandwidths, and accept impedances in a range appropriate to much biological work (though they are rarely suitable for four-terminal

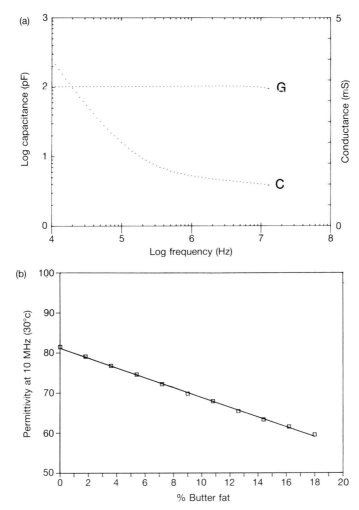

Figure 12. (a) Frequency-dependent impedance of a sample of single cream [butterfat content 18% (w/v)]. Measurements were made using a Hewlett-Packard model 4192A Impedance analyser and pin-type Pt black electrodes (41). The temperature was 30°C. All frequency-dependence of the impedance of this system is due to electrode polarization. (b) Effect of fat content on the capacitance of milk/cream mixtures. The cream [butterfat content 18% (w/v)] was progressively diluted with skimmed milk, and its impedance measured at 10 MHz in a thermostatted cell held at 30°C using a Hewlett-Packard 4192A Impedance Analyser and Pt black pin-type electrodes. The line is a least-squares fit and has a correlation coefficient of −1.00.

measurements). Our own experience with the model 4192 has permitted the acquisition of much data on the impedance of microbial suspensions in a time (before cell death) and with an ease that would have been inconceivable had a manually balanced bridge been used. Under computer control, the frequency, exciting voltage (current) and measurement time may be set, and the data acquired (over 6 decades at 20 frequencies per decade), stored, transformed, evaluated and plotted, at a rate of some 10 min per sample (41). We illustrate the utility of such a device by plotting the frequency-dependent capacitance and conductance of a sample of cream in *Figure 12a*. Because cream is an oil-in-water emulsion its dielectric properties (other than those due to electrode polarization effects) are frequency-independent in the range studied. The obvious convenience of using capacitance (permittivity) measurements to give a value for the fat content of milk/cream mixtures (on line and in real time) is illustrated in *Figure 12b*.

Spectrum Analysers of a type known as Dynamic Signal Analysers probably provide the most sophisticated (if expensive) approach to the measurement of impedances. The analogue current and voltage signals are coverted into digital form and transformed (usually via a Fourier or Laplace transformation) into a series of frequency bands, such that the Impedance $\mathbf{Z}(\omega)$ is given by

$$\mathbf{Z}(\omega) = [E(j\omega).I^*(j\omega)]/[I(j\omega).I^*(j\omega)] \tag{13}$$

where $I^*(j\omega)$ is the complex conjugate of $I(j\omega)$. Since each of these frequencies is separable (via Fourier's theorem and the hardwired algorithms usually built in to these devices), it is possible simultaneously to apply any exciting signal, which may be (and commonly is) a sum of sinusoids, an impulse function or pseudorandom binary noise. The chief and obvious merit of this approach, particularly using synthesized sinusoids, is that the benefits of sinusoidal excitation are combined with the multiplexing or multifrequency properties of time-domain analysis. Indeed (using the notation in *Figures 1* and *2*) the so-called transfer function of any linear impedance is:

$$\mathbf{Z}(\omega) = |Z(\omega)| \exp [j\theta(\omega)] \tag{14}$$

such that the modern usage of digital signal-processing devices exploiting this fact is beginning to revolutionize impedimetry generally.

4.3 Other measurement considerations

Apart from what has been stated above, and a general consideration of electronic noise, pickup and grounding and so on (42, 43), which are outside the scope of the present chapter, the chief source of experimental error in impedimetric and conductimetric devices is due to temperature fluctuations. Conductivity and permittivity have temperature coefficients of some $1-2\%$ K^{-1}, such that for sensitive and accurate work (where in some cases we are wishing to pick up changes in conductivity of some nS cm^{-1} on a background of mS cm^{-1}), temperature control of a high order is necessary (in this case clearly to 10^{-4} K or better). Evidently this requirement can be alleviated in some degree by time-averaging the signal, but temperature control is certainly a design consideration that commands a high priority for devices of this type.

We have here surveyed some of the general approaches to impedimetric devices.

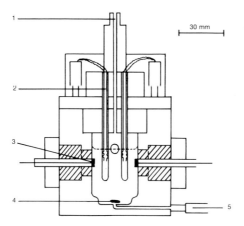

Figure 13. The conductimetric enzyme reaction cell devised by J. M. Wallach (personal communication). Buffer is introduced through port **5** via a peristaltic pump. The enzyme is introduced via port **1** and temperature equilibration carried out as judged by the temperature probes **2**. When the thermal drift has attained a constant value, based upon the conductivity measured with the Pt electrodes **3**, the substrate solution is introduced via **1** from a microsyringe. Solutions are stirred magnetically using the follower **4**. Figure courtesy of Prof. J. M. Wallach.

We therefore conclude with a survey of some of the types of conductimetric and impedimetric devices that have actually been built and used by a variety of workers for biosensing purposes, so as to leave the reader with the (accurate) view that conductimetry and impedimetry represent transducing principles of some maturity and promise in the biosensing world.

5. Implementation and commercial devices

5.1 Enzymatic

As pointed out, for instance, by Wallach (44) and by Lowe (45), many enzyme-catalysed reactions are linked by conductivity changes, so that in principle both enzymes and substrates may be assayed conductimetrically. Wallach and colleagues use a solution technique, rather than an immobilized enzyme, with the 4 ml reaction cell shown in *Figure 13*. The substrate is added and, after a thermal equilibration (with a temperature variation of $0.05°C$), the reaction is initiated with a small volume of the enzyme to be assayed. For serum butyrylcholinesterase, the required serum volume is $1-2$ μl, the conductance change some tens of μS per hour, and the response time some 2 min. The initial conductivity change is strictly proportional to the enzyme activity. The reaction cell is available from Tacussel, and whilst this device is perhaps not strictly a biosensor, it serves to illustrate the utility of conductimetry in enzyme and substrate assays.

Conductimetric and impedimetric devices

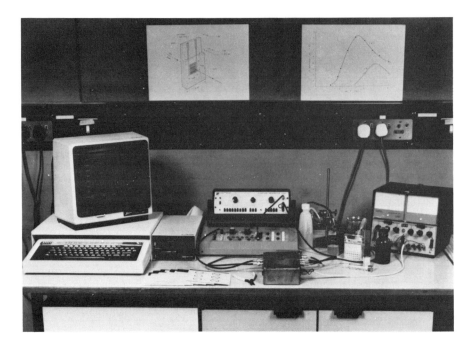

Figure 14. The Orbit Impedance Analyser, a device for measuring the frequency-dependent impedance of conductive media. Figure courtesy of Prof. R. Pethig, University College of North Wales.

Pethig and colleagues (46) have exploited interdigitated microelectrodes that they had previously used (47) in dielectrophoretic studies. A urease/bovine serum albumin (BSA) mixture was immobilized on the surface of these electrodes using glutaraldehyde, and the conductivity changes upon adding urea solutions assessed using an Orbit Impedance Analyser (*Figure 14*; Orbit Biotechnology Ltd). Urea solutions (in 150 mM NaCl) of concentrations as low as 1 mM gave conductance changes of some 0.5 mS/per minute. The response time was again under 2 min, although the range of linearity of rate versus [urea] was somewhat limited by the relatively high affinity of urease for its substrate. Given the high concentrations of urea normally encountered in clinical work, a simple dilution step would obviously solve this problem (whilst enhancing the precision). The chief advantage of this approach is that the confinement of the sensing area to the neighbourhood of the electrode surface maximizes the conductivity change for a given amount of enzyme, since the enzyme concentration is then high. This also has the benefit of lowering the apparent affinity of the enzyme and hence the domain of linearity.

Lowe and colleagues in collaboration with a group at Plessey (48) have taken a similar conductimetric approach even further. They have developed a microelectronic conductimetric biosensor which uses a dual-sensor configuration (one with and one

Figure 15. The Malthus AT microbiological conductance analyser. The 128 reaction vessels with their electrodes are held in the chamber on the left of the computer. Figure courtesy of Dr D. Millbank.

without enzyme) to increase the sensitivity and specificity. The electrodes are again of the interdigitated variety [see (41)], formed (using microelectronic fabrication techniques) of platinum on a silicon substrate. Urea was again the chief determinand, and the biosensing layer was a urease/BSA/glutaraldehyde film. The reproducibility was ±1%, and the device gave a correlation coefficient >0.99 when tested on 25-fold diluted samples against a standard method in a clinical laboratory.

5.2 Microorganisms

As with all living cells, microorganisms contain enzymes whose operation is linked to conductivity changes, and it has been known since the last century [see (4,49)] that the conductivity of culture broths changes according to the microbial load in a sample. Thus measuring time-dependent conductivity changes in a sample allows assessment of its microbial load (49,50). A number of commercially available devices are based on this principle (49), and one such, the Malthus Model AT Microbiological Analyser (Malthus Instruments) is illustrated in *Figure 15*. This is a fairly sophisticated, multiplexed device, controlled by a personal computer, in which 128 channels may be monitored simultaneously. The electrodes are of Pt on a ceramic

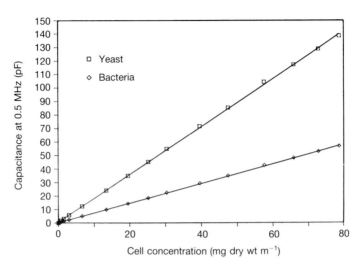

Figure 16. The dependence of the electrical capacitance of cellular systems upon the cell concentration. Measurements were carried out at a frequency of 500 kHz using the βUGMETER device and a four-terminal probe with a cell constant of 0.41 cm^{-1}. The βUGMETER was backed off to read 0 pF in the absence of cells. Yeast (*S.cerevisiae*) was obtained locally and suspended in 12 mM KCl. The bacteria were a strain of *Clostridium* sp. under study in this laboratory. The slopes of the two lines are essentially in proportion to the cell 'radii', as expected from theory (52).

substrate which is designed to fit a standard culture bottle. Temperature control is ensured by immersing the samples in a carefully thermostatted water bath. In this case, the conductimetric/impedimetric approach is probably the method of choice for assessing the sparse microbial load in difficult matrices such as foodstuffs (49).

Although the Malthus device, and cognate examples of this methodology, exploit the changes in impedance of culture broths and on the electrode−electrolyte interface caused by microbial growth and metabolism (51) in sparse loads, when the cell concentration is greater a direct assessment of microbial biomass is possible, for instance by measurement of the dielectric properties of fermentor broths *in situ* (52). This is because the β-dispersion (see above) exhibited by all cellular material generates a permittivity far greater than that of the suspending medium and in direct proportion to the cellular biomass (*Figure 16*). To this end, we have devised a novel instrument. The βUGMETER, in collaboration with Dulas Engineering, and ICI Biological Products, for the real-time estimation of microbial (and other) biomass (52). The instrument has a four-terminal configuration, for the reasons explained in Section 3.3, and gives an output of capacitance and conductance via a phase-sensitive detector. The capacitance signal may be calibrated to read in mg ml^{-1} biomass, and the measurement frequency may be set in the range 0.1−10 MHz. The instrument is illustrated in *Figure 17* and a typical experimental output is given in *Figure 18*. The output is for practical purposes instantaneous, and the sensitivity corresponds

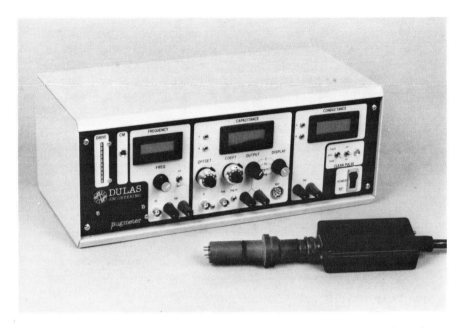

Figure 17. The Aber Instruments Model 212 βUGMETER, an instrument for the real-time estimation of biomass, and whose specification is described in the text.

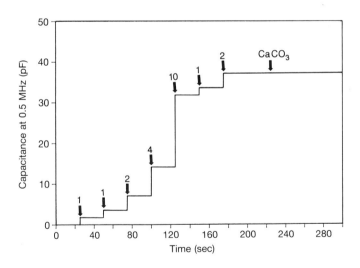

Figure 18. Typical output from the βUGMETER. Measurements were carried out as described in the legend to *Figure 17*. At the arrows, *S.cerevisiae* cells were added from a concentrated suspension to the concentrations (mg dry wt ml^{-1}) indicated. At the point marked CaCO$_3$ powdered calcium carbonate was added to a final concentration of 30 mg ml^{-1}.

Conductimetric and impedimetric devices

Figure 19. The experimental set-up used for magneto-inductance measurements by Blake-Coleman, Clarke and colleagues. Figure courtest of Dr B. C. Blake-Coleman.

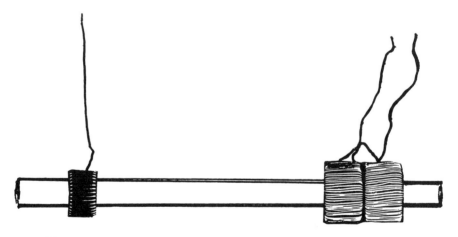

Figure 20. The principle of magneto-inductive measurements, in which two transformer coils are coupled to each other via the impedance to be measured. Figure courtesy of Dr B. C. Blake-Coleman.

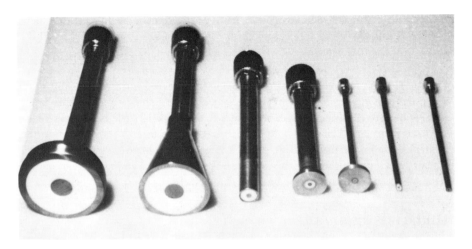

Figure 21. A selection of dielectric probes as used for work at frequencies between 10 kHz and 10 GHz by Gabriel, Grant and coworkers (54). Figure courtesy of Dr C. Gabriel.

(depending upon the electrodes used) to some 0.02 mg dry weight ml^{-1} for cells the size of *Saccharomyces cerevisiae*. Probe fouling is avoided by the use of automated electrolytic cleaning pulses. The instrument is available from Aber Instruments.

Although perhaps not strictly impedimetric device, a related approach has been exploited by Blake-Coleman, Clarke and colleagues (53) for the estimation of microbial biomass. The approach is referred to as magneto-inductance, and the experimental set-up is illustrated in *Figure 19*. The principle (*Figure 20*) is that the current induced in a secondary coil by a 'primary' coil depends upon the impedance of the intervening medium, such that measurements thereof, or of the resonant frequency of the system, provide a value for the (vector) impedance at the stated frequency.

5.3 Tissues

The impedance of living tissues depends upon a variety of factors (19), but is known to correlate with certain disease states. Gabrielli, Grant and colleagues (54) have developed an *in vivo* monitor for assessing the impedance of the skin and underlying tissues, using a number of probes illustrated in *Figure 21*. A time-domain measurement system is used, the frequency of measurement corresponding to the range 1 MHz – 10 GHz. The probes are connected (via an APC-7 connector) to a time-domain dielectric spectrometer, and consist of a precision air-line terminated by a ground plate and a 4 mm impedance-matched PTFE-filled section. Non-invasive estimation of the fat content of tissues represents an obvious application of this technology in biosensing. Since the impedance of tissues is well known to decrease monotonically (and in a frequency-dependent manner) following death, it is clear that the impedimetric approach based on probes of this type also represents a simple and non-invasive means of assessing (for instance) the time of death, a measurement of obvious importance in forensic studies.

6. Concluding remarks

Whilst the previous paragraph may have ended on something of a macabre note, we hope to have illustrated that conductimetric and impedimetric biosensors *sensu lato* have many areas of application, that several have already progressed to the commercial stage, and that there are grounds to believe that these are but the vanguard of a whole army of biosensing devices based on both linear and non-linear studies of the electrical organization of biological materials.

Acknowledgements

We are grateful to the SERC Biotechnology Directorate, ICI Biological Products and Aber Instruments for support of our work in this area, and to the Department of Trade and Industry for a SMART Award. We thank Drs B.C.Blake-Coleman, D.J.Clarke, C.Gabriel, D.Millbank, R.Pethig and J.M.Wallach for their kindness in supplying diagrams and photographs of their impedimetric instrumentation, and Anthony Pugh for photographic work.

References

1. Stock, J. T. (1984). *Anal. Chem.*, **56**, 561A.
2. Campbell, I. D. and Dwek, R. A. (1984). *Biological Spectroscopy*. Benjamin-Cummings, London.
3. Macdonald, J. R. (1987). *Impedance Spectroscopy*. John Wiley, New York.
4. Kell, D. B. (1987). In *Biosensors: Fundamentals and Applications*. Turner, A. P. F., Karube, I., and Wilson, G. S. (ed.), Oxford University Press, Oxford, p. 427.
5. Kell, D. B. and Harris, C. M. (1985). *J. Bioelectricity*, **4**, 317.
6. Cole, K. S. and Cole, R. H. (1941). *J. Chem. Phys.*, **9**, 341.
7. Schwan, H. P. (1957). *Adv. Biol. Med. Phys.*, **5**, 147.
8. Jonscher, A. K. (1983). *Dielectric Relaxation in Solids*. Chelsea Dielectrics Press, London.
9. Dissado, L. A. and Hill, R. M. (1983). *Proc. R. Soc. Lond. Ser. A*, **390**, 131.
10. Hill, T. L. (1977). *Free Energy Transduction in Biology*. Academic Press, New York.
11. Fersht, A. R. (1985). *Enzyme Structure and Mechanism*. 2nd edn, Freeman, San Francisco.
12. Kell, D. B., Astumian, R. D., and Westerhoff, H. V. (1988). *Ferroelectrics*, **86**, 59.
13. Barlow, D. J. and Thornton, J. M. (1986). *Biopolymers*, **25**, 1717.
14. Kirkwood, J. G. (1932). *J. Chem. Phys.*, **2**, 351.
15. Oncley, J. L. (1943). In *Proteins, Amino Acids and Peptides*. Cohn, E. J. and Edsall, J. T. (eds), Reinhold, New York, p. 543.

16. Shawhan, E. N. and Loveland, J. W. (1966). In *Encyclopedia of Industrial Chemical Analysis*. Snell, F. D. and Hilton, C. L. (ed.), Interscience, New York, p. 263.
17. Grant, E. H., Sheppard, R. J., and South, G. P. (1978). *Dielectric Behaviour of Biological Molecules in Solution*. Oxford University Press, Oxford.
18. Pethig, R. (1979). *Dielectric and Electronic Properties of Biological Materials*. Wiley, Chichester.
19. Pethig, R. and Kell, D. B. (1987). *Phys. Med. Biol.*, **32**, 933.
20. Kell, D. B. (1983). *Bioelectrochem. Bioenerg.*, **11**, 405.
21. Kell, D. B. and Harris, C. M. (1985). *Eur. Biophys. J.*, **12**, 181.
22. Harris, C. M. and Kell, D. B. (1985). *Eur. Biophys. J.*, **13**, 11.
23. Hanai, T. (1968). In *Emulsion Science*. Sherman, P. (ed.), Academic Press, London, p. 353.
24. Clausse, M. (1983). In *Encyclopedia of Emulsion Technology*. Becher, P. (ed.), Marcel Dekker, New York, Vol. 1, p. 481.
25. Schwan, H. P. (1963). In *Physical Techniques in Biological Research*. Nastuk, W. L. (ed.), Academic Press, New York, Vol. VIB, p. 363.
26. Cole, K. S. and Kishimoto, U. (1962). *Science*, **136**, 381.
27. Marmont, G. (1949). *J. Cell. Comp. Physiol.*, **34**, 351.
28. Geddes, L. A. (1972). *Electrodes and the Measurement of Biological Events*. Wiley, New York.
29. Grubbs, D. S. and Worley, D. S. (1983). *Med. Biol. Eng. Comput.*, **21**, 232.
30. Gielen, F. L. H. and Bergveld, P. (1982). *Med. Biol. Eng. Comput.*, **20**, 77.
31. Gottesfeld, S. and McIntyre, J. D. E. (1979). *J. Electrochem. Soc.*, **126**, 742.
32. Arwin, H., Lundström, I., and Palmqvist, A. (1982). *Med. Biol. Eng. Comput.*, **20**, 362.
33. Andrade, J. D. (1985). In *Surface and Interfacial Aspects of Biomedical Polymer, Volume 2, Protein Adsorption*. Andrade, J. D. (ed.), Plenum Press, New York, p. 1.
34. Ivarsson, B. A., Hegg, P.-O., Lundström, K. I., and Jönsson, U. (1985). *Colloids and Surfaces*, **13**, 169.
35. Schwan, H. P. and Ferris, C. D. (1968). *Rev. Sci. Instr.*, **39**, 481.
36. Tamamushi, R. and Takahashi, K. (1974). *J. Electroanal. Chem.*, **50**, 277.
37. Kibble, B. P. and Rayner, G. H. (1984). *Coaxial AC Bridges*. Adam Hilger, Bristol.
38. Furukawa, T., Nakajima, K., Koizumi, T., and Date, M. (1987). *Jap. J. Appl. Phys.*, **26**, 1039.
39. Westerhoff, H. V., Astumian, R. D., and Kell, D. B. (1988). *Ferroelectrics*, **86**, 79.
40. Gabrielli, C. (1980). *Identification of Electrochemical Processes by Frequency Response Analysis*. Solartron Electronic Group, Farnborough.
41. Harris, C. M. and Kell, D. B. (1983). *Bioelectrochem. Bioenerg.*, **15**, 11.
42. Morrison, R. (1977). *Grounding and Shielding Techniques in Instrumentation*. Wiley, New York.
43. Buckingham, M. J. (1983). *Noise in Electronic Devices and Systems*. Ellis Horwood, Chichester.
44. Wallach, J. M. (1983). *Progr. Clin. Enzymol.*, **2**, 317.
45. Lowe, C. R. (1985). *Biosensors*, **1**, 3.
46. Pethig, R., Lawton, B. A., Price, J. A. R., and Wray, T. A. K. (1988). In *Electrostatic Charge Migration*. Sproston, J. L. (ed.), Institute of Physics Short Meeting Series, IOP Publishing, Bristol, Vol 14, p. 21.
47. Price, J. A. R., Burt, J. P. H., and Pethig, R. (1988). *Biochim. Biophys. Acta*, **964**, 221.

48. Watson, L. D., Maynard, P., Cullen, D. C., Sethi, R. S., Brettle, J., and Lowe, C. R. (1987/88). *Biosensors,* **3**, 101.
49. Harris, C. M. and Kell, D. B. (1985). *Biosensors,* **1**, 17.
50. Richards, J. C. S., Jason, A. C., Hobbs, G., Gibson, D. M., and Christie, R. H. (1978). *J. Phys. E., Sci. Instr.,* **11**, 560.
51. Hause, L. L., Komorowski, R. A., and Gayon, F. (1981). *IEEE Trans. Biomed. Eng.,* **BME-28**, 403.
52. Harris, C. M., Todd, R. W., Bungard, S. J., Lovitt, R. W., Morris, J. G., and Kell, D. B. (1987). *Enz. Microb. Technol.,* **9**, 181.
53. Blake-Coleman, B. C., Calder, M. R., Carr, R. J. G., Moody, S. C., and Clarke, D. J. (1984) *Trends Anal. Chem.,* **3**, 229.
54. Cabriel, C., Grant, E. H., and Young, I. R. (1986). *J. Phys. E., Sci. Instr.,* **19**, 843.

6

Microbial biosensors

ISAO KARUBE and MASAYASU SUZUKI

1. Introduction

Recently, many kinds of biosensors have been developed for the determination of organic compounds (1–3). Enzyme sensors are highly specific for substrates of interest, but the enzymes employed are generally expensive and unstable. Microbial sensors are composed of immobilized microorganisms and an electrochemical device and are suitable for on-line control of biochemical processes.

The advantages of microbial biosensors are as follows (4):

- They are less sensitive to inhibition by solutes and are more tolerant of suboptimal pH and temperature values than enzyme electrodes.
- They have a longer lifetime than enzyme electrodes.
- They are cheaper because an active enzyme does not have to be isolated.

However, they also have disadvantages which are as follows:

- Some have a longer response time than enzyme electrodes.
- They need more time to return to the base line signal after use.
- Cells contain many enzymes and care has to be taken to ensure selectivity, e.g. by optimizing storage conditions or adding specific enzyme reactions. Mutant microorganisms lacking certain enzymes can be used.

The microbial sensors developed by our group involve the assimilation of organic compounds by microorganisms, changes in respiration activity, or the production of electrochemically active metabolites; the latter being monitored directly by an electrochemical device. This chapter describes procedures for the construction of and measurement with microbial biosensors.

2. Principles of microbial sensors

Microbial biosensors consist of a combination of microorganisms immobilized in a membrane with electrochemical devices. Microbial biosensors are classified either

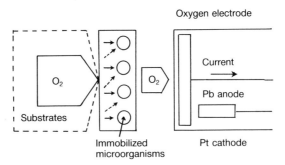

Figure 1. Respiration activity measurement type microbial biosensors.

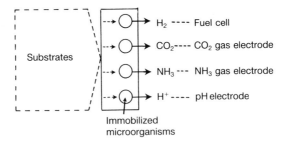

Figure 2. Metabolites measurement type microbial biosensors.

as respiration activity measurement type or electrochemically active metabolites measurement type.

In the case of respiration activity type, changes (normally, increases) in respiration activity of microorganisms caused by assimilation are detected by an oxygen electrode. From these changes, substrate concentrations are estimated. The principle of this type of microbial biosensor is shown in *Figure 1*. Aerobic microorganisms are used in these sensors. A microbial biosensor is dipped into buffer solution saturated with oxygen. Upon the addition of substrate, the respiration activity of the microorganisms is increased, which causes a decrease in oxygen concentration near the membrane. Using an oxygen electrode, substrate concentration can be measured from the oxygen decrease.

The other type of microbial biosensor detects electrochemically active metabolites such as H_2, CO_2, NH_3 and organic acids, which are secreted from microorganisms. The principle of these sensors is shown in *Figure 2*. This type of microbial biosensor is not limited to aerobes but can also employ anaerobic microorganisms.

Most of the microbial biosensors developed are of the respiration activity type.

I. Karube and M. Suzuki

3. Construction of microbial sensors

3.1 Immobilization of microorganisms

3.1.1 Immobilization techniques for biosensors

Many methods have been employed for immobilizing microorganisms. Most of these methods can be used for microbial biosensor construction; however, care must be taken with the following points.

- In the case of oxygen electrode-based sensors, gas permeability through the microorganism-immobilized membrane is very important.
- If it is based on the functions of living cells, a very gentle method for microbe-immobilization must be selected.

In early work on microbial biosensors, gel entrapment methods, e.g. polyacrylamide and collagen, have been employed. But nowadays, most microbial biosensors are constructed using membrane entrapment methods.

3.1.2 Membrane entrapment (5)

This is probably the mildest method of entrapment and is described below.

Protocol 1. Membrane entrapment of microbial cells

1. Culture the appropriate microorganism to a cell density corresponding to an absorbance at 562 nm of 8.6.
2. Take a porous cellulose acetate membrane and drop 3 ml of the microbial culture on to it whilst applying gentle suction from a water pump.
3. Air dry the membrane containing the entrapped cells and store it at room temperature.

Note: A Millipore Type HA membrane with a pore size of 0.45 μm, a diameter of 47 mm, and a thickness of 150 μm is suitable.

Either cellulose acetate or nitrocellulose membranes may be used. The method is illustrated in *Figure 3a(i)*. If vinyl spacers (14 mm o.d., 6 mm i.d., and thickness 50 μm) are attached to the cellulose acetate membrane as shown in *Figure 3a(ii)* several microbe-immobilized membranes may be made at one time. In *Figure 3a(iii)* the method of trapping cells between two membranes is shown; this technique has the advantage of making the membrane easier to handle and store. In all cases the membranes are kept in buffered saline.

Perhaps the easiest way to prepare microbe-immobilized membranes is illustrated in *Figure 3b*. The cells are first grown on an agar plate and then scraped off the agar with a spatula and further grown on the membrane.

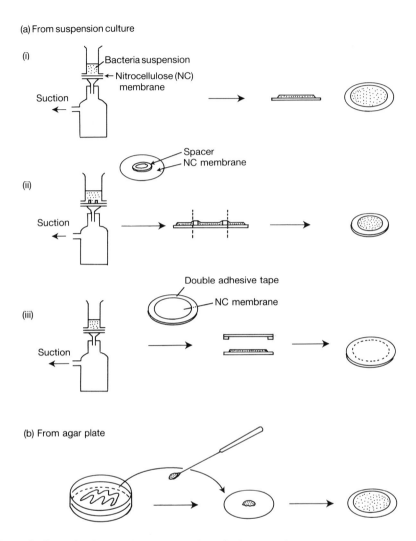

Figure 3. Examples for membrane entrapping of microorganisms.

3.1.3 Gel entrapping (6)

Microorganisms can be entrapped in a variety of gels but two that are particularly suitable for sensors are either polyacrylamide or agar. The methods for these are described below and a more general discussion of cell entrapment can be found in a companion volume in this series (*Immobilised Cells and Enzymes*).

Protocol 2. Polyacrylamide entrapment of cells

1. Grow the appropriate organisms and harvest the cells by centrifugation.

Protocol 2 *continued*

2. Wash the cells by resuspending them in physiological saline solution and then recentrifuge them.
3. Place a 20 mesh nylon net (1.3 cm diameter and 200 μm thick) on a clean glass plate.
4. Suspend 100 mg of the washed cells, 90 mg of acrylamide and 10 mg of N,N'-methylene bisacrylamide in 1 ml of physiological saline solution at 0°C.
5. Saturate the suspension of cells and acrylamide with nitrogen.
6. Initiate the polymerization by adding 30 μl of 10% (w/v) dimethylaminopropionitrile and 0.1 mg of potassium persulphate.
7. Cast 250 μl of the suspension over the nylon net and place the whole assembly in an oxygen-free atmosphere at 37°C for 30 min.
8. Remove the membrane and store in 0.1 M phosphate buffer at pH 7.

Protocol 3. Agar entrapment of microbial cells

1. Grow and wash the cells as described in steps 1 and 2 of the above protocol.
2. Suspend 100 mg of washed cells in 0.1 ml of physiological saline.
3. Dissolve 20 mg of agar in 0.9 ml of physiological saline at 100°C and then cool the solution to 50°C.
4. Mix the agar solution with the cell suspension and cast 250 μl of the resulting mixture on to a nylon net on a glass plate. Cool to 5°C.
5. After the agar has set, store the membrane in 0.1 M phosphate buffer pH 7.

3.2 Electrochemical devices for microbial biosensors

3.2.1 Oxygen electrode

A commercially available conventional dissolved oxygen electrode is the most general and suitable transducer for a microbial biosensor. The membrane electrode according to Clark is widely used (7). In the case of the Clark-type electrode, the geometrical configuration is very important. In particular the thickness of the electrolyte layer between the cathode and the membrane must be narrow so as to ensure good linearity and a low background current. Clark electrodes are classified as either polarographic electrodes or galvanic electrodes (*Figure 4*).

Polarographic electrodes consist of a platinum cathode and a silver anode both immersed in the same solution of saturated potassium chloride. A suitable polarization voltage between the anode and cathode selectively reduces oxygen at the cathode:

cathode: $O_2 + 2H_2O + 4e^- \rightarrow 4OH^-$
anode: $4Ag + 4Cl^- \rightarrow 4AgCl + 4e^-$

These chemical reactions result in a current which is proportional to the dissolved oxygen concentration.

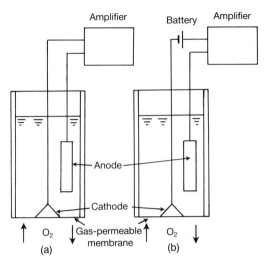

Figure 4. Principles of oxygen electrode. (a) Galvanic type. (b) Polarographic type.

The galvanic electrode has a lead anode and a silver (or platinum) cathode, and gives rise to a potential difference. Therefore it is a self driven electrode and does not need an externally supplied voltage. The reactions are as follows:

cathode: $O_2 + 2H_2O + 4e^- \rightarrow 4OH^-$
anode: $2Pb + 6OH^- \rightarrow 2PbO_2H^- + 2H_2O + 4e^-$

This type of electrode is very simple and economical, but it has disadvantages since it shows a slower response time and a shorter stability than a polarographic electrode.

3.2.2 Other electrodes

Fuel cell type electrodes (H_2-detection) (8), CO_2 electrodes (9), NH_3 electrodes (10), pH electrodes (including ISFETs) (11), can also be used for detectors in microbial biosensors. As most of these electrodes, except for the fuel cell type, are based on potentiometry, although they have a wide measurable range, they respond to other contaminants and have the limitation of lower detection limits.

3.2.3 Other possible devices for microbial biosensors

Many other physical devices can also be applied to microbial biosensors.

Thermistors (12)

If the immobilized microorganisms are placed in proximity to a thermistor which measures the metabolic heat evolved by them, microbial biosensors can be constructed (*Figure 5*).

As an example, yeast gel grains are packed in an insulated glass column provided with a thermistor at its outlet. The solution to be analysed is then pumped through a heat exchanger in an ultrastable thermostatically-controlled water bath and then

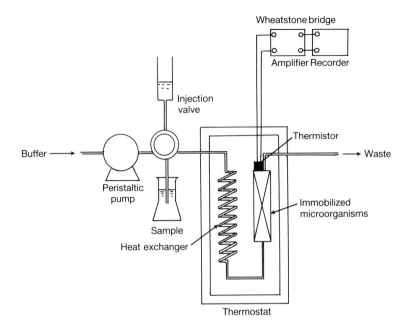

Figure 5. Microbial thermistor.

through the column. Heat changes occurring in the gel bed are recorded. Glucose, fructose and casein could be measured with this system. Additional examples of thermistor-based sensors are provided in Chapter 8.

Photo detectors (13)
Combining photobacteria with a photo detector (e.g. photomultiplier or photodiode), means that high sensitivity microbial biosensors can be constructed. Luminescence intensity of photobacteria (luminobacteria) is dependent on metabolic activity. Therefore nutrients of the microorganisms (e.g. glucose, amino acids) and inhibitors (e.g. toxicants, heavy metals) can be detected using this type of device. Generally luminescence intensity is a more sensitive parameter for metabolic activity than respiration activity or heat generation. Obviously only photobacteria can be used for this purpose.

3.3 Construction of microbial biosensors
3.3.1 Construction procedures
Culture of microorganisms. Culture procedures of microorganisms for microbial biosensors are the same as normal microbial culture. In general, log-phase growing cells are suitable for microbial biosensors.

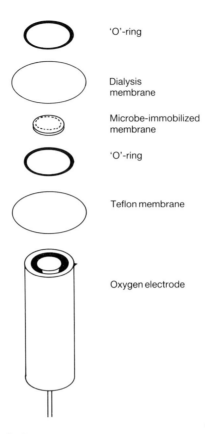

Figure 6. Construction of microbial biosensors.

Preparation of microbe-immobilized membrane. Microorganism-immobilized membrane is prepared by one of the methods mentioned in Section 3.1.

Construction of a microbial biosensor. The scheme of a microbial biosensor is illustrated in *Figure 6*. A Clark-type oxygen electrode consists of a Teflon membrane, a platinum cathode, an aluminium anode and saturated potassium chloride electrolyte. The microorganism-immobilized membrane is placed on the Teflon membrane and fixed in place using either a bored cap or a dialysis membrane/'O'-ring. The latter assembly is shown in *Figure 6*.

3.3.2 Measurement procedures

Batch measurement. Figure 7 shows a schematic diagram of the batch measurement system. The sensor is placed in a thermostatically-controlled, circulating water jacket. The temperature is maintained at 30°C, and the cell contains a total volume of 50 ml of 0.05 M phosphate buffer at pH 7.0. The buffer is saturated with dissolved oxygen (aeration rate: 200 ml min^{-1}) and stirred magnetically during the measurement.

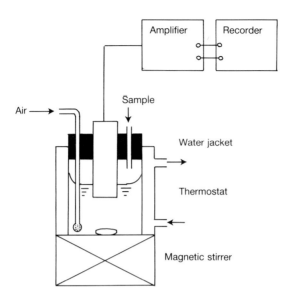

Figure 7. Batch measurement system.

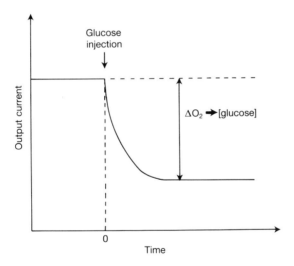

Figure 8. Typical response curve of microbial biosensors.

The current output of the oxygen electrode is measured with a digital multimeter (e.g. Takeda Riken, model TR6843) and with an electronic polyrecorder (e.g. TOA Electronics, model EPR-200A). When the output current becomes stable, 100 μl of sample solution is injected into the sensor cell. *Figure 8* shows a typical response

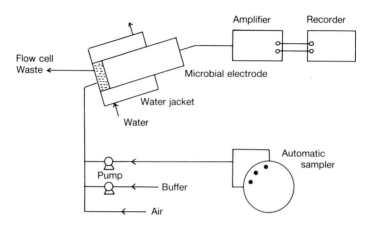

Figure 9. Flow measurement system.

curve of the electrode. The current at time zero is that obtained in a sample solution saturated with dissolved oxygen. The bacteria begin to utilize glucose in a sample solution when the electrode is placed in it. Then, consumption of oxygen by the bacteria causes a decrease in dissolved oxygen at the membrane. As a result, the current of the electrode markedly decreases with time until a steady state is obtained. The steady state indicates that the consumption rate of oxygen by the bacteria and the diffusion rate of oxygen from the solution to the membrane are equal. The steady state depends on the concentration of glucose.

When the electrode is removed from the sample and placed in a solution free of glucose, the current of the microbial electrode gradually increases and returns to the initial level.

Flow measurement. A flow system for a microbial biosensor consists of a jacketed flow cell containing a microbial electrode, a peristaltic pump, an automatic sampler, and a recorder (*Figure 9*). The temperature of the flow cell is maintained at 30°C by passing warm water through the jacket. Phosphate buffer solution (0.01 M, pH 7.0) saturated with dissolved oxygen is transferred to the flow cell at 1 ml min^{-1} together with air at a flow rate of 250 ml min^{-1}. When the electrode reaches a steady state, a sample is injected into the flow cell at a rate of 0.2 ml min^{-1} for 20 min. The sample is injected at 60 min intervals.

3.3.3 Maintenance

Generally microbial biosensors have a long lifetime but maintenance of these sensors is very important.

(a) Total activity of immobilized microorganisms should be kept almost constant. Normally, microbial biosensors should be stored in phosphate buffer without nutrients at 4 °C, otherwise microorganisms grow in the membrane. If the activity of the microorganisms decreases, the sensor must be dipped into nutrient medium for some time until the activity is recovered by growth of new cells.

(b) Contamination. Care to avoid contamination is much more important than for enzyme biosensors because antibacterial reagents, such as sodium azide or antibiotics, cannot be used. Contamination will affect selectivity and sensitivity of microbial biosensors.

4. Applications of microbial biosensors

4.1 Application fields of microbial biosensors

Many kinds of microbial biosensor have been developed and some of them have already been used in practical ways for environmental monitoring and in the food industry. Some of the microbial biosensors developed are summarized in *Table 1* (14).

4.2 BOD sensor (5)

The biochemical oxygen demand (BOD) is one of the most widely used tests in the management of organic pollution. The conventional BOD test requires, however, a 5 day incubation period. Therefore, a more rapid and reproducible method is required for assessing BOD. *Trichosporon cutaneum*, which is used for waste water treatment, is used for the BOD sensor. The sensor configuration is the same as described in Section 3.3. Phosphate buffer solution (0.01 M, pH 7) saturated with dissolved oxygen is transferred to the flow cell at a flow rate of 1 ml min^{-1}. When

Table 1. Microbial biosensors.

Sensor	Immobilized microorganisms	Device	Response time (min)	Range (mg dm^{-3})
Assimilable sugars	*Brevibacterium lactofermentum*	O_2-probe	10	10 – 200
Glucose	*Pseudomonas fluorescens*	O_2-probe	10	$2 - 2 \times 10$
Acetic acid	*Trichosporon brassicae*	O_2-probe	10	3 – 60
Ethanol	*Trichosporon brassicae*	O_2-probe	10	2 – 25
Methanol	Unidentified bacteria	O_2-probe	10	$5 - 2 \times 10$
Formic acid	*Citrobacter freundii*	Fuel cell	30	$10 - 10^3$
Methane	*Methylomonas flagellata*	O_2-probe	2	$0 - 6.6^a$
Glutamic acid	*Escherichia coli*	CO_2-probe	5	8 – 800
Cephalosporin	*Citrobacter freundii*	pH electrode	10	$10^2 - 5 \times 10^2$
BOD	*Trichosporon cutaneum*	O_2-probe	15	3 – 60
Lysine	*Escherichia coli*	CO_2-probe	5	$10 - 10^2$
Ammonia	Nitrifying bacteria	O_2-probe	10	0.05 – 1
Nitrogen dioxide	Nitrifying bacteria	O_2-probe	3	$0.51 - 255^b$
Nystatin	*Saccharomyces cerevisiae*	O_2-probe	1 h	$0.5 - 54^c$
Nicotinic acid	*Lactobacillus arabinosis*	pH electrode	1 h	$10^{-5} - 5$
Vitamin B_1	*Lactobacillus fermenti*	Fuel cell	6 h	$10^{-3} - 10^{-2}$
Cell population	–	Fuel cell	15	$10^8 - 10^{9d}$
Mutagen	*Bacillus subtilis* Rec$^-$	O_2-probe	1 h	$1.6 - 2.8 \times 10^3$

^{a} mmol, ^{b} p.p.m., ^{c} Unit cm^{-3}, ^{d} Number cm^{-3}

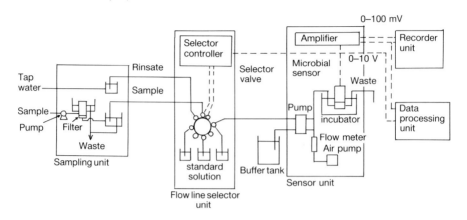

Figure 10. Commercial BOD sensor system.

the current reaches a steady-state value, a sample is injected into the flow cell at a rate of 0.2 ml min^{-1}. The steady-state current is dependent on the BOD of the sample solution. After the sample has been washed out the current of the microbial sensor gradually returns to the initial level. The response time of the microbial biosensor (time required for the current to reach a steady state) depends on the nature of the sample solution.

A linear relationship is observed between the current difference (between initial and final steady-state currents) and the 5-day BOD assay of the standard solution up to 60 mg l^{-1}. The minimum measurable BOD is 3 mg l^{-1}. The current is reproducible within 6% of the relative error when a BOD of 40 mg l^{-1} is employed over 10 experiments. The BOD sensor system shown in *Figure 10* has now been commercialized in Japan.

4.3 Gas sensors (15,16)

Gas sensing devices can also be constructed using immobilized microorganisms and an electrochemical device. Microbial biosensors for methane and carbon dioxide have been developed. A CO_2 sensor using an autotrophic bacterium is introduced here. A potentiometric CO_2 electrode has been developed and is commercially available, but various ions and some organic and inorganic volatile acids affect the potential of the inner pH electrode and the gas-permeable membrane of the CO_2 electrode. Also, precision of the conventional potentiometric CO_2 is somewhat restricted because of the Nernstian type of response. The autotrophic bacterium, *Pseudomonas* sp. S-17 (The Fermentation Research Institute, Japan), which can grow with only carbonate as the source of carbon, is used here. S-17 is autotrophically incubated under aerobic conditions at 30°C for approximately one month. The bacteria is immobilized on to the top of an oxygen electrode as described in Section 3.1. The area around the sensing selection is packed with a cell covered with a gas-

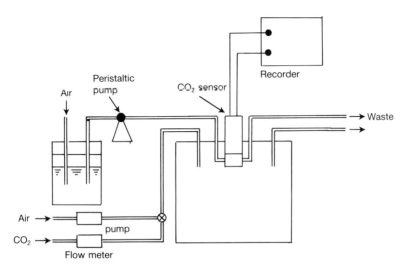

Figure 11. CO_2 gas sensor.

permeable membrane, a Millipore polytetrafluoroethylene (PTFE) membrane (pore size 0.5 μm, type FH) on one side. The cell is supplied with fresh, oxygen-saturated buffer solution (pH 6.5) containing some metal ions and 200 μM glucose. Experiments are performed in air using the apparatus shown in *Figure 11*. CO_2 is supplied to the container with the carrier gas (air). A linear relationship was obtained in a 3% to 12% CO_2 concentration in air. The sensor has a lifetime of more than a month.

4.4 Electrochemical bioassay (17)

The agar diffusion method is the conventional method for the bioassay of antibiotics, but it is not suitable for antibiotics that consist of heterogeneous mixtures of closely related compounds which are poorly soluble in water, and diffuse poorly in agar gel. Also, those antibiotics which are unstable in the presence of bright sunlight and produce zones of inhibition that may be neither clear nor proportional in size to the logarithm of the antibiotic's concentration are difficult to assay by this method. Here a novel bioassay for nystatin, using a microbial biosensor, is described. It is thought that nystatin binds to the sterol present in the membranes of sensitive cells leading to the formation of pores. The subsequent death of the microorganism is preceded by the leakage of cellular materials. Microbial death can be detected with an oxygen electrode. Using this principle, a yeast electrode composed of a membrane supporting immobilized yeast (*Saccharomyces cerevisiae*) cells is attached to an oxygen electrode and covered by a collagen membrane to prevent loss of the yeast cells. Nystatin is dissolved in dimethylformamide (DMF) and diluted to the appropriate concentration with 0.05 M phthalate buffer (pH 4.5) containing 500 mg l^{-1} of glucose. The

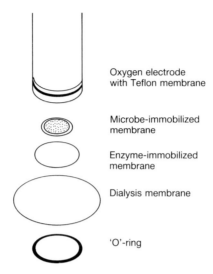

Figure 12. Construction of hybrid biosensors.

concentration of DMF in the sample solution must be kept below 0.66%. The yeast electrode is then inserted into the sample solution which is saturated with dissolved oxygen by stirring. Nystatin concentration can be measured from the rate of current increase as the dying cells consume less oxygen. Concentrations of greater than 0.5 unit ml^{-1} nystatin can be measured with this device.

4.5 Hybrid biosensor (18)

By combining cells with an immobilized enzyme membrane, one of the disadvantages of microbial biosensors, which is low selectivity, can be improved. As an example, a creatinine sensor is introduced here.

Creatinine deaminase (EC 3.5.4.21) hydrolyses creatinine to N-methylhydantoin and the ammonium ion, whence the ammonia produced is successively oxidized to nitrite and nitrate by nitrifying bacteria. The bacteria have not been completely characterized but are known to be a mixed culture of *Nitrosomonas* sp. and *Nitrobacter* sp. The sequence of reaction is as follows:

$$\text{Creatinine} + H_2O \xrightarrow{\text{creatinine deaminase}} NH_4^+ + N\text{-methylhydantoin}$$

$$NH_4^+ \xrightarrow{\textit{Nitrosomonas} \text{ sp.}} NO_2^- \xrightarrow{\textit{Nitrobacter} \text{ sp.}} NO_3^-$$

The bacteria consume oxygen during ammonia oxidation, so that the oxygen decrease may be detected with an oxygen electrode. The hybrid creatinine sensor thus consists of a cellulose dialysis membrane, immobilized creatinine deaminase, immobilized nitrifying bacteria, and an oxygen electrode, *Figure 12*. Construction of this electrode is described below.

Protocol 4. Construction of a hybrid biosensor

1. Suspend 300 mg (wet weight) of bacteria in 5 ml of sterile water.
2. Drip the suspension of cells on to a porous cellulose acetate membrane whilst applying slight suction from a water pump.
3. Rinse the membrane with 6 ml of 10 mM borate buffer pH 8.5.
4. Dissolve 250 mg of cellulose triacetate in 5 ml of dichloromethane and add 200 µl of 50% glutaraldehyde and 1 ml of 4-aminomethyl-1,8-diaminooctane to it. Spread the mixture on to a glass plate and allow the solvent to evaporate. Dry the membrane for 3−5 days and then cut it to the same size as the membrane containing the bacterial cells.
5. Immerse the membrane from step 4 in a 1% glutaraldehyde solution for 1 h and then rinse it with distilled water.
6. Place the glutaraldehyde activated membrane in a solution of 1 mg ml^{-1} creatinine deaminase (sp. act. 2.3 U mg^{-1}, Kyowa Hakko Kogyo Co.) dissolved in 10 mM phosphate buffer pH 7 for at least 15 h at 4°C.
7. Place the bacteria containing membrane over the Teflon membrane of an oxygen electrode so that the bacterial side contacts the Teflon. Put the enzyme membrane over the bacterial membrane and cover the entire assembly with dialysis membrane (Visking) held in place with an 'O' ring.

Note: A Millipore membrane such as that described in Section 3.1.2 is suitable for step 2.
The 1% glutaraldehyde should be freshly prepared from a 25% solution (Sigma Type I, kept at −20°C).

The analytical system consists of the sensor inserted in a flow cell, a carrier solution, a peristaltic pump (Model SJ-1211, Mitsumi Scientific Industry), an injection port, and a recorder (Model EPR 20A, TOA Electronics).

References

1. Turner, A. P. F., Karube, I., and Wilson, G. S. (ed.) (1987). *Biosensors: Fundamentals and Applications.* Oxford University Press, Oxford.
2. Mosbach, K. (ed.) (1988). *Methods in Enzymology.* Academic Press Inc., London and New York, Vol. 137.
3. Schmid, R. D. (ed.) (1987). *Biosensors International Workshop 1987.* VCH, Weinheim.
4. Wilson, K. and Goulding, K. H. (eds) (1986). *A Biologist's Guide to Principles and Techniques of Practical Biochemistry.* 3rd edition, Edward Arnold, London.

5. Hikuma, M., Suzuki, H., Yasuda, T., Karube, I., and Suzuki, S. (1979). *Eur. J. Appl. Microbiol. Biotechnol.*, **8**, 289.
6. Matsunaga, T., Karube, I., and Suzuki, S. (1980). *Eur. J. Appl. Microbiol. Biotechnol.*, **10**, 235.
7. Operating instructions O_2 electrodes, Ingold Co.
8. Matsunaga, T., Karube, I., and Suzuki, S. (1980). *Eur. J. Appl. Microbiol. Biotechnol.*, **10**, 235.
9. Hikuma, M., Obana, H., Yasuda, T., Karube, I., and Suzuki, S. (1980). *Anal. Chim. Acta*, **116**, 61.
10. Rechnitz, G. A., Kobos, R. K., Riechel, S. J., and Gebauer, C. R. (1977). *Anal. Chim. Acta*, **94**, 357.
11. Matsunaga, T., Karube, I., and Suzuki, S. (1977). *Anal. Chim. Acta*, **88**, 233.
12. Mattiasson, B., Larsson, P. O., and Mosbach, K. (1977). *Nature*, **268**, 519.
13. Lloyd, D., James, K., Williams, J., and Williams, N. (1981). *Anal. Biochem.*, **116**, 17.
14. Karube, I. (1987) In *Biosensors: Fundamental and Applications*. Turner, A. P. F., Karube, I., and Wilson, G. S. (ed.), Oxford University Press, Oxford, p. 13.
15. Suzuki, H., Tamiya, E., and Karube, I. (1987). *Anal. Chim. Acta*, **199**, 85.
16. Suzuki, H., Tamiya, E., Karube, I., and Oshima, T. (1988). *Anal. Lett.*, **21**, 1323.
17. Karube, I., Matsunaga, T., and Suzuki, S. (1979). *Anal. Chim. Acta*, **109**, 39.
18. Kubo, I., Karube, I., and Suzuki, S. (1983). *Anal. Chim. Acta*, **151**, 371.

7

Semiconductor field effect devices

FREDRIK WINQUIST and BENGT DANIELSSON

1. Introduction

The research and development of biosensors is at present in a rapidly growing phase due to the demands from a number of different areas such as clinical chemistry, process industry, agriculture and environmental control. Many different types of biosensor have been developed, and the majority of the biosensor concepts are based on the combination of immobilized enzymes with classical sensors such as photometers, amperometric or potentiometric electrodes, gas electrodes or thermistors. The performance of a biosensor is to a large extent dependent on the sensor elements. One class of sensors that has been developed during the last decade is the solid state chemical sensors; these have properties that make them especially valuable for biosensing purposes. Based on semiconductor technology, they are characterized by their ruggedness, ability to be directly integrated with microelectronics and hence the possibility to be mass fabricated at low cost. One type of such sensor is based on field effect devices, made by the MOS (Metal Oxide Semiconductor) technology. They can be designed either as capacitors or field effect transistors.

ISFETs (Ion Sensitive Field Effect Transistors) were introduced in 1970, and were the first type of this class of sensor in which a chemically sensitive layer was integrated with solid state electronics (1). By excluding the gate metal in an FET and using a pH sensitive gate insulator, a pH sensitive ISFET could be constructed. It was discovered in 1975 that a large hydrogen gas sensitivity for MOS structures could be obtained if the gate material was made of the catalytically active metal palladium (2). Since then, several types of MOS structures with catalytically active gates have been fabricated and investigated and it was recently found that the ammonia gas sensitivity of this type of device could be considerably increased if a very thin catalytic metal, such as platinum or iridium (with a nominal thickness of $2-40$ nm), is used as the gate (3). It was also found that these TMOS (Thin Metal film Oxide Semiconductor) structures could be made very sensitive to unsaturated hydrocarbons and alcohols if operated at high temperatures ($>170°C$) (4,5).

In this chapter, a brief survey of the physics and response characteristics of field effect devices is described and discussed as well as fabrication techniques for ISFET, PdMOS and TMOS devices. Bioanalytic systems involving ISFETs as well as hydrogen and ammonia gas sensitive MOS structures in combination with enzymes will also be described.

These kinds of bioanalytical systems represent a class of biosensors which are in a rapid state of progress. Most of the systems described in this chapter are thus not yet commercialized but are still in the experimental stage.

2. Basic principles of field effect devices

2.1 A brief survey of semiconductor physics

A semiconductor is characterized by the limited amount of free charge carriers. The total concentration of free charge carriers can, however, be increased by impurity doping. Trivalent impurities, such as boron, in a lattice of the tetravalent semiconductor silicon will take up an electron from the surroundings, and the immobile impurity atoms will thus be negatively charged and the surroundings will have an excess of positive mobile charges. The silicon is then p-type. Correspondingly, pentavalent dopant atoms, such as phosphorus, in a lattice of silicon will result in positive immobile ions and negative mobile charges (electrons), and the silicon is then n-type. The mobile free charge carriers are affected by the electrical field, which makes it possible to control the concentration and behaviour of these charge carriers by external means.

MOS field effect devices can be constructed either as capacitors or field effect transistors (FET), as depicted in *Figures 1a and 2a*, respectively. If an electrical field is applied over the gate insulator (the oxide) of the devices, the number of mobile charges at the surface of the semiconductor is changed. In an MOS capacitor, the capacitance of the device will change, and in an FET the current along the semiconductor surface will change. This is outlined in *Figure 1b* where the capacitance of an MOS device versus the applied voltage is shown, and in *Figure 2b* where the current along the surface of an FET (I_{DS}) versus the gate voltage (V_G) is shown.

One basis for the sensor function of these devices is that if charged or polar species (ions) adsorb between the gate and the insulator, then a corresponding charge will add at the semiconductor surface. As a result, the capacitance–voltage curve of an MOS capacitor and the I_{DS} versus V_G curve for an FET will shift along the voltage axis as outlined in the marked lines in *Figures 1b* and *2b*, respectively. Depending on the sign of the charged species, the curves will shift either to the left or to the right. The voltage shift thus obtained is then a measure of the concentration of adsorbed species.

In research, field effect devices in the form of capacitors are most often used, since they are very easy to fabricate. Although field effect transistors are more complicated to fabricate, they have, however, advantages which make them especially valuable. The electronic equipment necessary is less complicated than that for capacitors; furthermore, the FET represents a qualitative development from common sensors, since the first amplification step is integrated within the sensor. This will result in a low impedance signal from the FET, which is virtually immune to electromagnetic disturbances. These points are discussed further in reference 6.

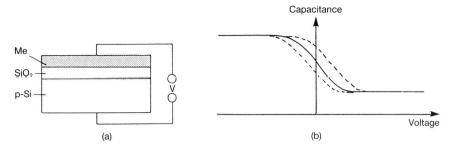

Figure 1. (a) Schematics of an MOS capacitor. (b) The corresponding capacitance–voltage curve. The marked lines show the parallel shift of the curve caused by adsorbed charged species.

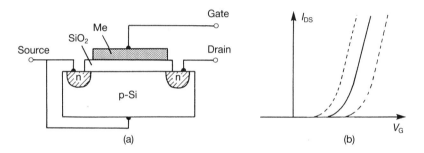

Figure 2. (a) Schematics of an MOS field effect transistor. (b) The corresponding drain current (I_{DS}) versus gate potential (V_G) curve. The marked lines show the parallel shift of the curve caused by adsorbed charged species.

2.2 Ion selective field effect transistors—ISFETs

An ISFET is constructed as a field effect transistor, with the metal gate replaced by an ion selective membrane, a solution and a reference electrode, as shown schematically in *Figure 3*. The current across the FET depends on the charge density at the semiconductor surface, which is controlled by the electric field of the semiconductor surface. This is in turn determined by the reference electrode and by ions interacting with the ion sensitive membrane. In principle, the response characteristic of the ISFET should follow the well known Nernst relation, that is, a logarithmic relationship is obtained between the change in gate potential, ΔV, and the change in analyte concentration, ΔC, according to equation 1:

$$\Delta V = RT \ln(\Delta C) \tag{1}$$

or, at $T = 300°K$ and with ΔV expressed in mV as shown in equation 2:

$$\Delta V = 58.6 \log(\Delta C) \tag{2}$$

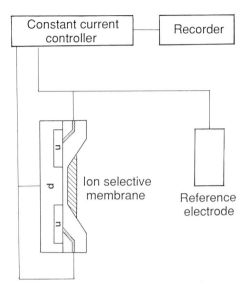

Figure 3. Schematics of an ISFET connected to a constant current controller and a reference electrode.

Thus, for a monovalent ion, a 10-fold increase in analyte concentration should ideally result in a change of the gate potential of 58.6 mV. In practice, however, it is observed that few ISFETs really obey this law; in the fabrication of the devices a large spread in response characters can be observed, even within batches. Due to improved fabrication techniques, however, the performance of the sensors is continually being improved. The major problems that hamper more extended use of ISFETs are the long-term drift in the baseline of the devices and the difficulty in achieving a good encapsulation; furthermore, the adhesion of the ion sensitive membrane is often poor. Due to the small distances between conducting areas on the surface of the sensor and its use in strong electrolytes, a current leakage path often develops. In practice, very few encapsulation materials are suitable. Often epoxy based resins are used.

2.3 Gas sensitive metal oxide semiconductor structures

In a hydrogen gas sensitive MOS structure the gate material consists of the catalytically active metal palladium. When hydrogen molecules in the ambient surroundings are adsorbed on the palladium surface, they dissociate into atoms. Some of these will diffuse through the palladium layer and adsorb on the metal−oxide interface where they are polarized, thus forming a charged dipole layer. On the surface of the palladium, a back reaction also takes place, in which adsorbed hydrogen atoms are oxidized and removed. This is essential to discharge the sensor when hydrogen no longer is present. These reactions are depicted in *Figure 4*. In order to keep water

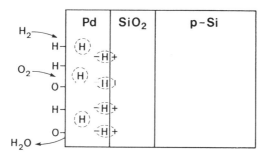

Figure 4. Chemical reactions occurring on the surface of a PdMOS sensor. For further details see text.

molecules off the surface of the palladium and to speed up the chemical reactions occurring in the devices, they are normally operated at temperatures up to 100–150°C. These structures can be constructed either as palladium gated capacitors (PdMOS cap) or field effect transistors (PdMOSFET). The voltage drop, ΔV, created by a hydrogen dipole layer will shift the $I_{DS}-V_G$ characteristics of a transistor and the $C-V$ characteristics of a capacitor to the left along the voltage axis, as shown in *Figures 1b* and *2b*, respectively. The voltage shift in a transistor is normally measured as the change in the voltage needed to keep a constant, pre-set current, and in a capacitor, as the change in the voltage needed to keep a constant, pre-set capacitance. The sensitivity to hydrogen in air is better than 1 p.p.m.V (parts per million by volume).

Ammonia gas sensitive MOS structures can be made if the gate oxide in the structure is covered by a thin (2–40 nm) porous layer of a catalytically active metal such as iridium or platinum. The mechanism causing the ammonia sensitivity of these devices is complicated and still not fully elucidated in detail, but it is caused by catalytic decomposition reactions of ammonia on the metal. As a result of these reactions, charged reaction intermediates are produced which will give rise to a surface potential change, which in turn is capacitively coupled through the pores in the metal film on the semiconductor surface. For the determination of ammonia, a thin layer of iridium may be used as the gate material. Such IrTMOS (Iridium Thin Metal film Oxide Semiconductor) structures are very selective for ammonia (only hydrogen gas and some low molecular weight amines appear to interfere) with a sensitivity of at least 1 p.p.m.V NH_3 in air.

The response characteristics of gas sensitive MOS structure are in principle determined by the type of catalytic reactions occurring at the surface of the metal. For the hydrogen sensor, the correlation of the voltage drop, ΔV, with the hydrogen gas partial pressure, P_{H_2}, follows equation 3:

$$\Delta V = \Delta V_{max} [(\alpha \cdot \sqrt{P_{H2}}/(1 + \alpha \cdot \sqrt{P_{H2}})] \qquad (3)$$

ΔV_{max} is the maximum obtainable voltage shift and α is a constant. At small concentrations of P_{H_2} equation 4 holds:

$$\Delta V = K_1 \cdot (P_{H_2})^{1/2} \qquad (4)$$

The value of the constant K_1 might vary from PdMOS structures between different batches, but a typical value is $20 [mV \cdot (p.p.m.V\ H_2)^{-1/2}]$ in air.

For the ammonia sensitive TMOS structures, the situation is more complicated due to the complex reactions ammonia molecules may undergo on the surface of the sensor. If, however, the sensors are exposed to ammonia at a low concentration during a short time, Δt, then the relationship is as in equation 5:

$$\Delta V = K_2 \cdot P_{NH_3} \cdot \Delta t \qquad (5)$$

where K_2 is a constant. This equation also applies to PdMOS structures. It is thus convenient to use short measurement pulses which will yield a linear relationship between the measured signal and the analyte concentration. This is utilized in the flow systems described in Section 2.4. More information on ISFET, PdMOS and TMOS structures can be found in refs 6–8, which also cover the basic theory of these devices.

2.4 Principal differences between ISFETs and gas sensitive MOS devices

Surface oriented gas sensors obey in principle a saturated response similar to equation 3, whereas ISFETs follow the Nernst relationship (equation 1). It can be perceived from these equations that the dynamic range for gas sensors is smaller than that of the ISFETs. In certain concentration regions, however, the gas sensors may have a larger differential sensitivity.

To use gas sensors in aqueous surroundings, gas-permeable membranes can be utilized for a phase separation step which offers certain advantages: the sensor will be electrically separated from the solution; furthermore, there is no need for a reference electrode. In addition, the measured compound will be present in a relatively higher concentration in the gas phase than in the liquid phase since the sensor can be placed in a comparatively small gas volume; this will make the sensor system very sensitive.

3. Experimental

3.1 Sensor fabrication and instrumentation

ISFETs are most often made as n-channel field effect transistors with open gate areas. The ion selective membrane is then deposited over the gate area by different techniques, such as chemical vapour deposition (CVD). For pH sensitive ISFETs, silicon nitride is most commonly used as the material for the ion selective membrane.

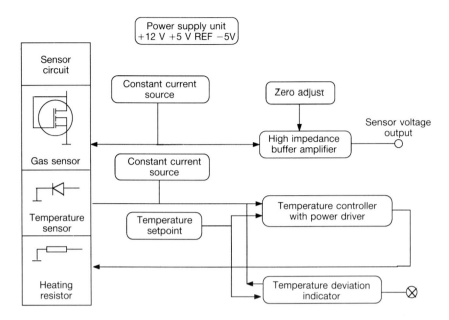

Figure 5. A schematic circuit diagram of a constant current controller.

The ISFET is thereafter encapsulated by a polymer (e.g. an epoxy based resin) leaving the gate area open. To effect measurements, the ISFET is connected to a reference electrode and a constant current controller, as shown in *Figure 3*. The potential change across the ion selective membrane due to the change in hydrogen ions in the solution is then measured as the change in voltage needed to keep a constant, pre-set current. A schematic circuit diagram of a constant current controller is shown in *Figure 5*. So far, ISFETs are not commercially available although there have been large efforts from a number of companies to introduce them on the market.

The IrTMOS capacitors used are fabricated in the same way. A p-type silicon wafer (100 orientation, Wacker-Chemitronic) with a resistivity of 5 Ω cm is cleaned in NH_4OH/H_2O_2 solution, each at 80°C for 5 min. Silicon dioxide is then grown to a thickness of 100 nm by dry thermal oxidation at 1200°C, whereafter the silicon wafer is allowed to cool in argon. Ohmic contact to the back of the wafer is made by etching away the silicon dioxide with hydrofluoric acid, followed by evaporation of aluminium to a nominal thickness of 100 nm. The silicon wafer is then annealed at 500°C for 10 min in an ambient atmosphere of 1% H_2 in N_2. Contact strips of palladium are thermally evaporated through Y-shaped masks (length 7 mm) to a nominal thickness of 200 nm. Iridium is then electron beam evaporated through a mask with circular holes (diameter 2 mm) over the upper parts of the Y-shaped palladium contacts. The nominal thicknesses of the iridium films are 4 nm. Finally, gold contacts are attached on the lower parts of the palladium contacts, and the sensors

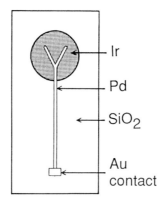

Figure 6. The IrTMOS capacitor. The size of the sensor is 5 × 10mm².

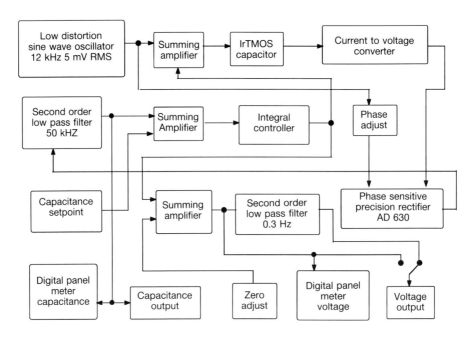

Figure 7. A schematic circuit diagram of a constant capacitance controller.

diced from the silicon wafer. From a 2 inch silicon wafer, 38 IrTMOS capacitors can be obtained. The final capacitor is shown in *Figure 6*. When used in the bioanalytical systems described in the following sections, the sensor is mounted on a temperature-controlled sample holder, normally adjusted to 35°C. The shift of the capacitance–voltage characteristics of the capacitor due to ammonia gas exposure

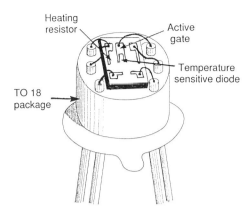

Figure 8. A PdMOSFET chip, including an integrated heater and a temperature sensing diode. The chip is mounted on a TO 18 holder.

is measured with a constant capacitance controller, operated at 25 kHz. The schematics of a circuit diagram of the constant capacitance controller is shown in *Figure 7*.

Hydrogen gas sensitive structures used in the biosensor systems are all of the PdMOSFET type. This sensor-chip also contains an integrated heater and a temperature sensitive diode, and the size of the chip is 1×1 mm^2. The sensor-chip is mounted on a TO 18 holder as shown in *Figure 8* and normally operated at 130°C. The catalytically active gate consists of about 100 nm evaporated palladium. The shift in the threshold voltage due to hydrogen gas exposure is measured as the change in the voltage needed to keep a constant, pre-set current. The schematics of a circuit diagram of the constant current controller is shown in *Figure 5*. IrTMOS capacitors and PdMOSFETs, including the electronics, can also be obtained from Sensor AB.

3.2 Preparation of enzymes

All enzymes described in the following section are commercially available, except for the enzyme hydrogen dehydrogenase (HDH, EC 1.12.1.2). Furthermore, this enzyme has a poor stability and must therefore be prepared in the laboratory. When immobilized, however, the stability of the enzyme is considerably improved. The methodology for the preparation and purification of HDH from the microorganism *Alcaligenes eutrophus H16* is given below.

Protocol 1. Preparation of HDH from *Alcaligenes eutrophus H16*

1. The microorganism can be obtained from the American Type Culture Collection, No. 17699.
2. Prepare 7 l of nutrient broth consisting of 3 g l^{-1} beef extract, 5 g l^{-1} peptone and 0.244 μM NiSO$_4 \cdot$6H$_2$O. Adjust the pH to 6.8.

Protocol 1 *continued*

3. Grow the microorganism in a 10 litre fermentor at 30°C and with a stirring rate of 80 rev min^{-1}, and bubble with gas mixture, containing 80% H_2, 10% O_2 and 10% CO_2 at 10 ml min^{-1} through the fermentor.

4. At the end of the growth phase, collect the cells and wash them with 50 mM sodium phosphate buffer at pH 7.0. The cells can be stored at $-30°C$ for several months.

5. During the enzyme preparation, all the work should be performed at $+4°C$. Disrupt the cell suspension by sonication for 15 min and centrifuge the preparation at 140 000 g for 120 min.

6. Add dropwise 21 ml of 2.5% w/v Cetavlon to each 100 ml of the supernatant. Remove the precipitate by centrifugation at 5000 g for 5 min.

7. Fractionate the preparation by adding solid ammonium sulphate and collect the fraction that precipitates between 40% and 60% saturation by centrifugation at 5000 g for 5 min.

8. Dissolve the precipitate in 0.02 M potassium phosphate buffer, pH 7.0, and dialyse for 12 h against 0.02 M potassium phosphate buffer, pH 7.0.

9. Pre-equilibrate 200 ml of swollen DEAE Sepharose CL6B (Pharmacia) with 0.02 M potassium phosphate, pH 7.0, and pack it in a 2.5 cm $\times$ 50 cm column.

10. Apply 5 ml of the crude enzyme preparation to the top of the column and elute with a total volume of 600 ml of a linear potassium chloride gradient from 0 to 1 M in 0.02 M potassium phosphate, at a flow rate of 0.4 ml min^{-1}. Collect 3-ml fractions.

11. Assay the fractions by measuring the initial rate of NAD^+ reduction spectrophotometrically in 1 cm pathlength cuvettes at 340 nm, using 50 mM Tris$-$HCl buffer at pH 8.0, containing 0.8 mM NAD^+ and 0.2 mM NADH, saturated with hydrogen gas.

12. Pool the most active fractions. Normally, the enzyme activity in these fractions should be at least 1 unit mg^{-1}.

A more detailed description of the properties of the enzyme can be found in ref. 9.

3.3 Immobilization of enzymes

Many properties of immobilized enzymes are different from those of the free enzyme. Important ones are the increased long-term and temperature stability; furthermore, free enzyme catalyses a homogeneous reaction, whereas immobilized enzymes are involved in heterogeneous catalysis, thus diffusion-controlled reaction kinetics may be important. A change in the pH profile and the Michaelis constant, K_M, is also often observed. A large variety of different methods and carriers materials is available for enzyme immobilization (10). The methods used in this study are based on covalent coupling to controlled pore glass (11), or to oxirane acrylic beads (12), as described below.

Protocol 2. Immobilization of enzymes to CPG (controlled pore gas)
1. Aminopropyl substituted CPG, mean pore diameter 200 nm, 80–120 mesh, is obtained from Corning Glass Works.
2. Activate 1 ml of wet, suspended CPG by the addition of 10 ml of glutardialdehyde (2.5% in 0.1 M sodium phosphate buffer, pH 7.0) and shake gently for 1 h.
3. Wash the preparation with two 10 ml aliquots of distilled water and two of the sodium phosphate buffer used in step 2.
4. Dissolve 20 mg of enzyme in 4 ml of the sodium phosphate buffer used in step 2 and add the CPG suspension. Shake gently at +4°C for 12 h.
5. Wash the immobilized enzyme preparation twice with 10 ml aliquots of the sodium phosphate buffer used in step 2.
6. Shake the suspension with 4 ml of 2 mM glycine in the same sodium phosphate buffer for 2h.
7. Repeat step 5.
8. Wash the enzyme preparation with two 5 ml aliquots of 0.5 M sodium chloride in the same sodium phosphate buffer.
9. Repeat step 5.

Protocol 3. Immobilization of enzyme to oxirane acrylic beads
1. Oxirane acrylic beads, with the trade name Eupergit C, are obtained from Röhm-Pharma GmbH.
2. Dissolve 5 mg of enzyme in 5 ml of 0.05 M sodium phosphate buffer, pH 7.0, and add 125 mg of dry Eupergit C to this solution.
3. Shake the mixture gently at room temperature for 36 h.
4. Wash the enzyme preparation with two 10 ml aliquots of the sodium phosphate buffer used in step 2.
5. Shake the suspension with 4 ml of 1 mM glycine in the same sodium phosphate buffer for 12 h.
6. Repeat step 4.
7. Wash the enzyme preparation with two 5 ml aliquots of 0.5 M sodium chloride in the same sodium phosphate buffer.
8. Repeat step 4.

In the ISFET study, the enzyme was directly attached to the pH sensitive surface according to the method described below.

Protocol 4. Immobilization of β-lactamase to an ISFET
1. Dissolve about 1 mg each of β-lactamase and bovin serum albumin in 100 μl of distilled water.

Semiconductor devices

Protocol 4 *continued*

2. Apply 2−5 μl of the protein solution over the active surface of the ISFET. Let the protein solution dry in air.
3. Apply 5 μl of 10% (by volume) glutardialdehyde in 0.05 M sodium phosphate buffer, pH 7.0, over the protein layer and incubate for 5 min.
4. Wash with the sodium phosphate buffer used in step 3.
5. Encapsulate the active surface of the sensor with a dialysis membrane.

3.4 Biosensing systems

Biosensor systems may vary from a simple enzyme-bound membrane wrapped around a pH electrode to large analytical instruments. Two main configurations can, however, be distinguished: enzyme electrode probes, more commonly denoted as bioprobes, and flow systems. The bioprobe is a simple and small device intended for direct measurements in solution. It often consists of an enzyme membrane in direct contact with the transducer device. These sensors are simple to make and easy to handle. Some disadvantages are, however, the limited amount of enzyme that can be attached, which makes the bioprobe sensitive to inhibitors and lowers its lifetime; it is also often difficult to obtain baseline accuracy. Furthermore, the bioprobe cannot usually be calibrated during the measurement. In the other system, which is based on flow injection analysis techniques, buffer flows through a reaction column containing immobilized enzyme and then to the transducer. Samples are injected prior to the reaction column. These systems are more complicated than bioprobes since the system contains a pump, a sample injector, an enzyme reactor and a suitable transducer. Additionally, these systems cannot be used for direct measurement in solution. Flow through systems have, however, features that make them especially valuable in bioanalysis: the reaction conditions in the enzyme column can be chosen so that the enzyme operates under optimal conditions with respect to parameters such as pH or temperature. Furthermore, the reaction column, if large enough, can contain a large capacity of enzyme activity. For further reading, see refs 13 and 14.

The ISFET-based biosensing system described in Section 4.1 is constructed as a bioprobe, while the biosensing systems based on PdMOS and IrTMOS devices are constructed as flow through systems.

3.4.1 Experimental arrangements using PdMOSFETs

A flow through system for the determination of NADH using immobilized HDH and a PdMOSFET sensor is shown schematically in *Figure 9*. A multichannel peristaltic pump is used to pump buffer continuously at a flow rate of 0.5 ml min^{-1} and carrier gas (air) at a flow rate of 2.5 ml min^{-1} through the system. The two flow streams enter the enzyme reactor. The mixture of buffer and carrier gas (now containing enzymatically produced hydrogen gas) from the outlet of the reaction column then enters a cell with a circular (20 mm diameter) gas-permeable membrane made of porous polytetrafluoroethylene (Fluoropore, Millipore), mean pore size

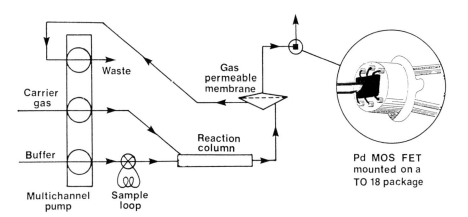

Figure 9. Experimental set-up for the determination of NADH and NAD$^+$ with a PdMOSFET. The reaction column contains immobilized hydrogen dehydrogenase.

0.2 μm. The carrier gas penetrates the membrane and is led to the PdMOSFET operated at 130°C. The multichannel pump is also used to pump out the aqueous phase from the gas separation cell. A sample injection valve with a 0.5 ml sample loop is inserted in the buffer flow stream, prior to the mixing chamber.

3.4.2 Experimental arrangement using ammonia gas sensitive IrTMOS structures

The flow through system for ammonia determinations in aqueous solution is based on gas diffusion across a gas-permeable membrane, as for the flow through system, based on the PdMOSFET. IrTMOS structures are, however, operated at much lower temperatures (around 40°C), so the sensor can be placed very close to the membrane due to the small temperature gradient. The system will thus be faster and the ammonia sensitivity larger. Furthermore, operation at low temperatures results in the electronic properties of the device being improved, thus the baseline will be more stable and the background noise level of the device may be considerably decreased.

The flow injection analysis system for ammonia is shown in *Figure 10*, omitting the reaction column. A peristaltic pump is used to pump buffer continuously through the system at a flow rate of 0.125 ml min^{-1} via a sample injection valve with a 85 μl sample loop to an ammonia sensing probe. This is constructed as a flow through cell (5 mm o.d., height 15 mm) which is made of Teflon and has an inner, circular cavity (3 mm diameter, 0.2 mm deep) at the edge which is connected with an inlet and an outlet for the buffer solution. A gas-permeable membrane (Plastlon, Garlock Plastomer Products, mean pore size 0.45 μm) and a Teflon gasket are mounted over the cavity. The flow through cell is pressed against the active surface of an IrTMOS structure via a spring. In some applications, the pump is also used to introduce alkaline buffer (pH 12.5) into the flow stream prior to the inlet to the ammonia sensing probe, which is indicated as a marked line in *Figure 10*.

Semiconductor devices

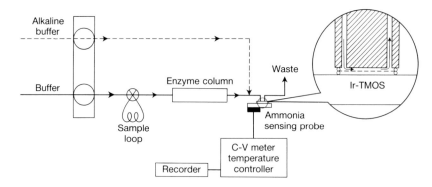

Figure 10. Experimental set-up for the determination of substrates to ammonia producing enzymes. In some applications, the flow stream is mixed with alkaline buffer, as shown in the marked line.

4. Methods

4.1 Bioanalytical uses of pH sensitive ISFETs

pH sensitive ISFETs can be used for the determination of any substrate that can enzymatically produce or consume protons. The enzyme can either be directly immobilized on the pH sensitive part of the sensor or on a membrane which is then attached to the sensor. The size of the sensing part of an ISFET is normally extremely small, thus only minute amounts of enzyme are necessary. This also means that substrate consumption is negligible, even in very small sample volumes.

4.1.1 Determination of penicillin using a pH sensitive ISFET

In this section, a method for the determination of penicillin is described. It is based on the action of the enzyme β-lactamase which catalyses the reaction where penicillin is hydrolysed and generates protons. The enzyme is directly immobilized on the pH sensitive part of an ISFET according to the third protocol in section 3.3. The sensor is set up as shown in *Figure 3* and calibrated as described below.

Protocol 5. Calibration of the penicillin sensitive ISFET

1. Connect the sensor to a constant current controller (set point 0.1 mA) and a reference electrode.
2. Place the reference electrode and sensor in a stirred solution of 0.05 M phosphate buffer pH 7.
3. Allow the signal from the sensor to stabilize.
4. Add known quantities of penicillin to the buffer and measure the change in equilibrium potential.

Protocol 5 *continued*

5. Construct a calibration curve by plotting the change in equilibrium potential against the penicillin concentration. The curve should follow equation 6.

$$\text{Concentration of penicillin (mM)} = K_{\text{Pen}} \cdot \Delta V \text{ (mV)} \quad (6)$$

The constant K_{Pen} should be in the order of 0.25. Normally, the detection limit for penicillin is at least 0.1 mM and the calibration curve is linear up to 25–50 mM. This set-up can now be used for assays of penicillin in solutions with the same buffer capacities as those used for the calibration.

4.2 PdMOSFET in bioanalysis

The enzyme HDH, immobilized in a reaction column and inserted in the flow through system shown in *Figure 9*, can be used for direct determination of NADH, since it catalyses the reaction:

$$NAD(P)H + H^+ \rightleftharpoons NAD(P)^+ + H_2$$

4.2.1 Determinations of NADH and NAD$^+$

Arrange an experimental set-up as described in Section 3.4.1 and shown in *Figure 9*. The reaction column, made of Teflon, should contain about 0.5 ml of HDH, immobilized to controlled pore glass with an apparent activity of about 5 U.

Protocol 6. Calibration of NADH

1. Prepare a solution of 0.5 M Tris buffer adjusted to pH 8 with hydrochloric acid.
2. Pass this buffer through the immobilized enzyme column at a flow rate of 0.5 ml min^{-1}.
3. Inject 0.5 ml of standard solutions of NADH into the buffer stream. The NADH concentration should be in the range 0.01–5 mM.
4. Measure the magnitude of the potential pulse (ΔV) in mV in the output of the PdMOSFET during the sample pulse.
5. Plot ΔV against the NADH concentration and the resulting curve should follow equation 7.

$$\text{Concentration of NADH (mM)} = K_{\text{NADH}} \cdot \Delta V \text{ (mV)} \quad (7)$$

A typical value of K_{NADH} is 0.01. The lower limit of detection for NADH is of the order of 0.03 mM and the analysis time for one sample is of the order of 6 min. The principle for the determination of NAD$^+$ is based on the reversibility of the

Semiconductor devices

hydrogenase catalysed reaction in which hydrogen gas is consumed. Use the same experimental set-up as for the determination of NADH.

Protocol 7. Calibration of NAD^+

1. Prepare a solution of 0.5 M Tris buffer adjusted to pH 8 with hydrochloric acid and containing 0.05 M NADH.
2. Pass this solution through the immobilized enzyme reactor at a flow rate of 0.5 ml min^{-1}.
3. Mix hydrogen with the air carrier gas to a volume concentration of 100 p.p.m.
4. Allow the output of the sensor to stabilize and then inject 0.5 ml aliquots of standard NAD^+ solutions into the buffer stream. The NAD^+ concentration should be in the range 0.05–0.6 mM.
5. There will be a change in potential output of the PdMOSFET in the opposite direction to that seen with NADH and the relationship between NAD^+ concentration and potential change (in mV) should follow equation 8.

$$\text{Concentration of } NAD^+ \text{ (mM)} = -K_{NAD^+} \cdot \Delta V \text{ (mV)} \qquad (8)$$

A typical value of K_{NAD^+} is 0.031, and the detection limit for NAD^+ is normally 0.05 mM.

4.3 Ammonia sensitive IrTMOS in bioanalysis

Ammonium is a weak acid with pK_a = 9.25. Ammonium ions therefore are in equilibrium with ammonia in the pH range 6–12, depicted in the reaction below:

$$NH_4^+ + OH^- \rightleftharpoons NH_3 + H_2O$$

The pH of the solution is thus an important parameter, since ammonia gas formation will be favoured by a raised pH and the sensitivity of the system will be correspondingly increased. The maximum sensitivity is obtained at pH values above 11. The flow through system can be used for the determination of ammonia in a wide variety of samples such as tap, river and rainwater, as well as biological fluids such as whole blood and serum. The blood specimens are analysed at pH 8.0, but for the inorganic samples a modified set-up is used (as indicated with the marked line in *Figure 10*) where the flow stream is mixed with a strong alkaline buffer at pH 12.5. The lower limit of detection is better than 0.3 μM at pH 8.0 and 0.1 μM at pH 12.0. One sample can be analysed in less than 3 min.

By combining reaction columns containing immobilized ammonia liberating enzymes with the system analysis of the substrates to these enzymes can be carried out. The performance of the system is influenced by the pH-dependent parameters: enzyme activity, enzyme stability and ammonia formation. The pH of the working buffer is normally chosen near the pH optimum of the enzyme activity. For very low pH values, however, the ratio of ammonia to ammonium ions will be very small, which will limit the sensitivity of the system. In such cases, the alternative flow through system is used by which the operating pH can be chosen to be the optimum pH of the enzyme, whereas the pH in the sample cell may be chosen to be considerably larger.

4.3.1 Determination of urea

For the determination of urea, the enzyme urease (EC 3.5.1.5) is used; it catalyses the reaction:

$$(NH_2)_2CO + H_2O \rightarrow 2NH_3 + CO_2$$

Protocol 8. Experimental set up and method for the determination of urea

1. Immobilize urease on Eupergit C as described in the second protocol in section 3.3 and pack the support into a 200 μl column.
2. Assemble the flow system shown in *Figure 10* and described in Section 3.4.2.
3. Prepare a solution of 0.05 M Tris buffer adjusted to pH 8.5 with hydrochloric acid. [Although the pH of the buffer is higher than the optimum for urease (7.3) the high enzyme loading (≈ 300 Units) and low substrate concentration ensures complete conversion of the urea to ammonia.]
4. Inject samples of urea into the buffer stream and measure the change in potential (in mV) as a function of the urea concentration. This latter should be in the range $1-50$ μM and the relationship between the two quantities should follow that of equation 9.

$$\text{Concentration of urea } (\mu M) = K_{\text{Urea}} \cdot \Delta V \text{ (mV)} \qquad (9)$$

where a typical value of K_{Urea} is 0.36. This set-up can now be used for clinical assays of urea in whole blood and serum. The samples should then be diluted 500-fold prior to analysis. One sample can be assayed for urea in less than 3 min.

4.3.2 Determination of creatinine

The assay method described below is based on the action of the enzyme creatinine iminohydrolase (EC 3.5.4.21) which catalyses:

$$\text{Creatinine} + H_2O \rightarrow N\text{-methylhydantoin} + NH_3$$

The normal endogenous background of ammonia in biological samples, such as whole blood or urine, is in the same range or higher than that of creatinine and is thus a serious interferent which must be removed. This is carried out by the enzymatic amination of α-ketoglutarate, catalysed by immobilized glutamate dehydrogenase (EC 1.4.1.3) according to the scheme:

$$NH_4^+ + NADH + \alpha\text{-ketoglutarate} \rightarrow \text{Glutamate} + NAD^+$$

Protocol 9. Experimental set up and method for the determination of creatinine

1. Immobilize creatinine iminohydrolase on Eupergit C as described in *Protocol 3*.
2. Immobilize glutamate dehydrogenase on to controlled pore glass as described in *Protocol 2* and pack 100 µl into a second column. This corresponds to an apparent activity of 7 Units. The glutamate dehydrogenase column is fitted upstream of the creatinine iminohydrolase column.
3. Insert the two columns into the flow system as shown in *Figure 10*.
4. Prepare a solution of 0.05 M Tris buffer adjusted to pH 8.5 with hydrochloric acid and containing 1.5 mM NADH and 0.5 mM α-ketoglutaric acid.
5. Inject 85 µl of standard solutions of creatinine into the flowing stream and measure the potential change (in mV) for concentrations of creatinine from 1 to 40 µM.
6. Plot the change in potential against the creatinine concentration; the curve should follow equation 10.

$$\text{Concentration of creatinine } (\mu M) = K_{Creatinine} \cdot \Delta V \text{ (mV)} \qquad (10)$$

where a typical value of $K_{Creatinine}$ is 0.71. The flow through system can be used for the determination of creatinine in whole blood and plasma after a 30-fold dilution, and in urine after a 1000-fold dilution.

4.3.3 Other nitrogen containing compounds

There are many ammonia producing enzymes acting on a large number of different substrates described in the literature. Many of these enzymes are also commercially available at reasonable cost, which means that measuring systems for their substrates can be simply arranged if based on the flow through system. In *Table 1*, a summary of investigated enzyme—substrate combinations is shown. The enzymes were in all cases immobilized on Eupergit C and the calibration curve for 85 µl standard solution could be expressed as:

$$\text{Substrate concentration } (\mu M) = K_{Sub} \cdot \Delta V \text{ (mV)} \qquad (11)$$

where K_{Sub} is a constant, dependent on the operating conditions of the system.

5. Summary

In this chapter, various biosensor systems based on field effect devices have been described as well as some practical measurements. Due to their early stage of development, these systems have not yet progressed beyond the research laboratories. The research concerning these structures is, however, at present very intense and they offer a new concept that will no doubt have a profound impact in the near future.

Table 1. Enzyme – substrate combinations.

Substrate	Enzyme	pH-optimum for the enzyme	pH of the working buffer	K_{Sub} $\mu M\,mV^{-1}$	Notes
Adenosine	Adenosine deaminase	7.5	8.5	0.71	High enzyme activity applied
Alanine	Alanine dehydrogenase	10	10	0.48	–
AMP	AMP-deaminase	6.5	6.5	0.46	Alternative system used; pH of the sample cell 12.5
Asparagine	Asparaginase	8.6	8.5	0.72	–
Creatinine	Creatinine iminohydrolase	7.9	8.5	0.71	The enzyme has a broad pH-optimum
Glutamate	Glutamate dehydrogenase	7.5	8.5	0.73	High enzyme activity applied
Glutamine	Glutaminase	4.9	4.9	0.48	Alternative system used; pH of the sample cell 12.5
Histidine	Histidine ammonia lyase	9.0	9.0	0.83	Low enzyme activity applied
Tryptophane	Tryptophanase	8.3	8.3	0.99	Low enzyme activity applied
Urea	Urease	7.3	8.5	0.36	Very high enzyme activity applied

Acknowledgements

We thank Professor Ingemar Lundström for valuable discussions and Ms Eva Hedborg for helping to prepare the manuscript. Financial support from the National Swedish Board for Technical Development is gratefully acknowledged.

References

1. Bervgeld, P. (1970). *IEEE Trans.*, **BME-17**, 70.
2. Lundström, I., Shivaraman, M. S., and Svensson, C. (1975). *Appl. Phys. Lett.*, **26**, 55.
3. Winquist, F., Spetz, A., Armgarth, M., Nylander, C., and Lundström, I. (1983). *Appl. Phys. Lett.*, **43**, 839.
4. Winquist, F. and Lundström, I. (1987). *Sensors and Actuators*, **12**, 255.
5. Ackelid, U., Winquist, F., and Lundström, I. (1986). *Proc. 2nd Int. Meet. Chemical Sensors*, Bordeaux, University of Bordeaux, France, p. 387.
6. Janata, J. and Huber, R. J. (ed.) (1985). *Solid State Chemical Sensors*. Academic Press, New York.
7. Lundström, I., Armgarth, M., Spetz, A., and Winquist, F. (1986). *Sensors and Actuators*, **10**, 399.
8. Sibbald, A. (1986). *J. Mol. Electronics*, **2**, 51.
9. Schneider, K. and Schlegel, H. G. (1976). *Biochim. Biophys. Acta*, **66**, 435.
10. Woodward, J. (ed.) (1985). *Immobilized Cells and Enzymes*. IRL Press, Oxford.
11. Weetall, H. H. and Filbert, A. M. (1974). In *Methods in Enzymology*, Jacoby, W. D. and Wilchek, M. (ed.), Academic Press, New York, Vol. 44, p. 134.
12. Hannibal-Friedrich, O., Chun, H., and Sernetz, M. (1980). *Biotechnol. Bioeng.*, **22**, 157.
13. Guilbault, G. (1984). *Analytical Uses of Immobilized Enzymes*. Marcel Dekker Inc., New York.
14. Turner, A., Karube, I., and Wilson, G. (eds) (1987). *Biosensors: Fundamentals and Applications*. Oxford University Press.

8

Thermometric sensors

BENGT DANIELSSON and FREDRIK WINQUIST

1. Introduction

Bioanalytical calorimetry has many attractive features, which were recognized early in studies using conventional calorimeters. Since biological reactions are usually more or less exothermic, calorimetry offers a general detection method which is insensitive to the optical properties of the sample. Enzymic reactions are associated with rather high molar enthalpy changes in the range of $20-100$ kJ mol^{-1} (*Table 1*) and it is often possible to base measurements on only one enzymic step. This is in contrast to other techniques where the detection is based on, for instance, the change in concentration of coloured reactants. In such cases it is usually necessary to couple the primary reaction with one or several (enzymic) reactions in order to obtain measurable changes.

The microcalorimeters used in these early biochemical studies were, however, rather sophisticated and expensive instruments with relatively slow response times, and therefore not suitable for rapid routine analyses (1). Over 10 years ago several different simple calorimetric devices were introduced which combined the general detection principle of calorimetry with the specificity of immobilized enzymes (2). Additional advantages included the re-usability of the biocatalyst, the possibility to work with continuous-flow systems, the insensitivity to the optical properties of the sample, and simple procedures. Several of the concepts introduced at that time have

Table 1. Molar enthalpies of some enzyme catalysed reactions.

Enzyme	Substrate	$-\Delta H$ (kJ mol^{-1})
Catalase	Hydrogen peroxide	100
Cholesterol oxidase	Cholesterol	53
Glucose oxidase	Glucose	80
Hexokinase	Glucose + ATP	28 (75)[a]
Lactate dehydrogenase	Na-pyruvate	62
NADH-dehydrogenase	NADH	225
β-Lactamase	Penicillin G	67 (115)[a]
Trypsin	Benzoyl-L-arginineamide	29
Urease	Urea (phosphate buffer, pH 7.5)	61
Uricase	Urate	49

[a] The ΔH values in parenthesis include protonation of Tris (-47.5 kJ mol^{-1}).

been used over the years in numerous applications. One of these devices is the Enzyme Thermistor (ET), which was designed in our laboratory (3) and is the main focus of this chapter.

2. Different principal designs of thermal biosensors

In connection with the more widespread introduction of immobilized enzymes in bioanalysis, several research groups attempted to develop simple, low cost calorimeters to be used for routine analyses, especially in clinical applications for the determination of metabolites such as glucose and urea. The most straightforward approach used thermal enzyme probes (TEP), in which the enzyme is directly attached to the temperature transducer, a thermistor, by either cross-linking or entrapping the enzyme in a dialysis bag which encloses the thermistor (4,5). With this arrangement, however, a major portion of the heat evolved in the enzymic reaction is lost to the surroundings without being detected by the thermistor. Consequently, the sensitivity is low, even though it is somewhat improved in subsequent designs (6,7). The TEP concept is primarily intended for batch operation similar to the 'small volume calorimeter' proposed at the same time. In this approach the enzyme is attached to thin aluminium foil placed on the surface of a Peltier element acting as a temperature sensor (8). The sample is applied as a small drop on the enzyme layer, and the amount of substrate is detected as a very small temperature change due to the enzymic reaction.

A considerably more efficient method of detecting the reaction heat is possible in systems employing a small column with the enzyme immobilized on a suitable support, as in the 'enzyme thermistor' (3,4) and in the 'immobilized enzyme flow-enthalpimetric analyser' (9), as well as in other similar techniques (10). The combination of a commercial flow-enthalpimeter with a thermostatted, immobilized enzyme column has also been described (11). In these cases the heat is transferred by the liquid passing through the column to or along the temperature sensor that is mounted at either the top of the column or at its outlet.

Recent developments of thermal biosensors include an integrated-circuit biocalorimetric sensor for glucose with total dimensions of only $1 \times 1 \times 0.3$ mm (12) and a thermoelectric glucose sensor employing a thin film thermopile to measure the evolved heat (13).

3. Instrumentation

3.1 A simple Plexiglas apparatus

Our initial studies on biothermal analysers were carried out with various simple Plexiglas housings for the immobilized enzyme column. These devices are thermostatted in accurately-controlled water baths, and the temperature at either the top of or at the outlet of the column is monitored with a small thermistor probe connected to a commercial Wheatstone bridge constructed for temperature measurements

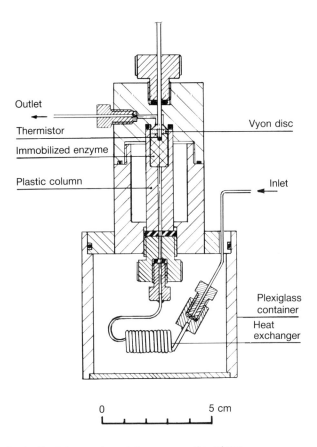

Figure 1. A simple Plexiglas version of the enzyme thermistor.

(Knauer Temperature Measuring Instrument). At its most sensitive setting the recorder output of this instrument is 100 mV for a temperature change of 0.01 °C. We also use bridges of our own construction with similar or higher sensitivity.

The simple device shown in *Figure 1* has been shown to be very useful for the determination of concentrations as low as 0.01 mM. This sensitivity is adequate for the determination of urea in 10-fold diluted serum samples (14). This dilution factor is high enough to eliminate problems with non-specific heat.

The plastic column, which can hold up to 1 ml of the immobilized enzyme preparation, is mounted with a insulating airspace around it. The heat exchanger coil (~50 cm of acid-proof steel tubing, 0.8 – 1 mm i.d.) is placed in a water-filled cup in order to minimize the influence of temperature fluctuations in the water bath (Heto Type 02 PT623 UO; temperature stability better than 0.01 °C). The temperature probe is made by attaching a suitable glass-encapsulated thermistor (e.g. Veco Type

41A28, 10 K at 25°C, 1.5 × 6 mm) with epoxy resin to the tip of a 2 mm (o.d.) acid-proof steel tube.

A buffer solution is pumped through the system by a peristaltic pump with low pulsation (LKB Varioperpex pump or Minipulse) at a flow rate of 1 ml min^{-1}. Samples (0.1–1 ml) are introduced using a three-way valve or a chromatographic sample valve. The height of the resulting temperature peak is normally used as a measure of the substrate concentration and can be linear with concentration over wide ranges, typically 0.01–100 mM. Up to 30 samples can be measured per hour.

3.2 Present enzyme thermistor design with aluminium calorimeter

To allow for more sensitive determinations and for more convenient operation we have subsequently developed a two-channel instrument in which the water bath is replaced by a carefully thermostatted metal block as shown in *Figure 2*. The calorimeter part of the instrument, which is placed in a container insulated with polyurethane foam, consists of an outer aluminium cylinder (80 × 250 mm) thermostatted at 25, 30 or 37°C, with a stability of at least ±0.01°C. Inside this jacket is a second aluminium cylinder with two column ports and a pocket for a reference thermistor. Before entering the column, the solution to be analysed is passed through thin-walled acid-proof steel tubing (0.8 mm i.d.) with two-thirds of its length in close contact with the thermostatted jacket. The last third is in close contact with the inner cylinder, which otherwise is thermally insulated from the outer one and acts as a heat sink. Short-term temperature fluctuations in the column will be exceedingly small through this arrangement. The columns are attached to the end of the Delrin tubes which are inserted into the calorimeter. These tubes also contain the temperature sensors (Veco Type A395 isocurve thermistors; 16 K at 25°C, temperature coefficient −3.9% per °), which are mounted with heat conducting epoxy on a short piece of gold capillary through which the column effluent passes. The Wheatstone bridge is of a DC-type and is built with precision resistors with a low temperature coefficient and a chopper-stabilized operational amplifier (MP 221 from Analogic Corp). The maximum sensitivity of this bridge is 100 mV per 0.001°C, but the most commonly used measurement ranges are 0.01–0.05°C. A major limiting factor of temperature resolution in practical use is temperature fluctuations due to friction and turbulence in the column, further accentuated by pulsations from the pump, which often prevents the use of higher sensitivities than 100 mV per 0.005°C. It should be noted that a reaction enthalpy change of 80 kJ mol^{-1} should give a temperature change of 0.01°C for 1 ml of a 1 mM sample at a flow rate of 1 ml min^{-1}. These are rather typical parameters for this type of analysis.

A number of units of the instrument described here have been produced at the workshop of our institute over the last 10 years for use in various university and industrial laboratories. The commercial sale of this and some other calorimetric instruments is now handled by Thermometric Co. The enzyme thermistor apparatus sold by Thermometric is called Thermal Assay Probe.

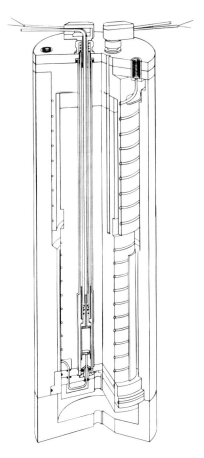

Figure 2. Schematic cross-section of an enzyme thermistor with an aluminium constant temperature jacket. There are two identical column ports (one temperature probe with a column is shown in cross-section) that can be used idependently or one of the ports can be used as a reference channel (split-flow).

3.3 A miniaturized enzyme thermistor

In order to see how far a functioning system for flow injection analysis can be scaled down our present studies involve, as an intermediary step, investigations of a miniaturized single-channel version of the instrument described in Section 3.2. In this version the length-scale has been reduced to 1/5, which means that the diameters of the flow channels are $0.1-0.2$ mm and the flow rate is $25-50$ μl min^{-1} with sample volumes of $1-10$ μl. The column, which has dimensions reduced to 1×3 mm, has to be packed with material having a particle size less than 0.1 mm. The performance of this device is comparable with that of the apparatus described in Section 3.1, which makes it suitable as a portable monitor, for instance in health care for the detection of metabolites present in concentrations exceeding 1 mM. Our current results indicate that the dimensions can be further reduced without a significant loss of performance.

4. Procedures

The procedures described here refer primarily to the metal block thermostatted instrument described in Section 3.2. As with the simple Plexiglas device, an essentially pulse-free pump (usually a good peristaltic pump) is employed to produce a continuous flow of buffer (0.5–5 ml min^{-1}) through the channels in use. Since water has a very high heat of vaporization, it is important that the apparatus be handled so as to prevent spillage inside the calorimeter and to avoid air bubbles from entering the system. The use of a pulse damper, capable of trapping incoming gas bubbles, is advisable and the solutions should be partially degassed, in an ultrasonic bath under vacuum for instance, at least in work done at higher sensitivities.

Samples are usually introduced with an injection valve (Type 50 from Rheodyne) using 0.1–0.5 ml sample loops. If sample volumes less than 25 μl are employed undiluted samples may be injected, but for larger sample volumes the samples should be diluted at least 5- to 10-fold in order to avoid non-specific heat effects from dilution or solvation. Such effects can also be eliminated by a split-flow technique (15), in which the flow is divided into two equal streams, one through the enzyme column and the other through a similar, but inactive, column. Since the temperature is measured differentially between the columns, non-specific heat effects should cancel. At high flow rates and with small sample volumes, one sample per minute can be analysed.

Columns of varying size holding up to 1 ml (at 7 mm i.d.) can be accommodated in the present apparatus. The enzyme support normally used is CPG (controlled pore glass) which is available in many different pore and particle sizes and with different ligands for coupling. In most cases we used propylamino-derivatized CPG with a pore size in the range of 500–2000 Å and a particle size in the order of 80 mesh (0.18 mm), with a large excess of enzyme, often 100 units or more, immobilized with glutardialdehyde. CPG offers high binding capacity, good mechanical and chemical as well as microbial stability and relatively simple coupling procedures (see the first protocol of Chapter 7; Section 3.3). Other support materials that have been used include Sepharose CL-6B (Pharmacia-LKB) and oxirane acrylic beads (Eupergit C; Röhm Pharma). For a summary of the procedure for binding enzymes to Eupergit C see the second protocol of Section 3.3, Chapter 7. Sepharose columns can only be used at flow rates below 0.5 ml min^{-1} or they will rapidly become compressed. They have mostly been used for antibody binding in the TELISA procedure described in Section 5.8.

The major limiting factor for column life appears to be mechanical obstruction.

Table 2. Substances that have been analysed with thermal biosensors.

Substance	Immobilized enzyme(s)	Concentration range (mM)
Ascorbic acid	Ascorbate oxidase	0.05 – 0.6
ATP	Apyrase	1 – 8
Cellobiose	β-Glucosidase + glucose oxidase/catalase	0.05 – 5
Cephalosporin	Cephalosporinase	0.005 – 10
Cholesterol	Cholesterol oxidase + catalase	0.03 – 0.15
Creatinine	Creatinine iminohydrolase	0.01 – 10
Ethanol	Alcohol oxidase/catalase	0.01 – 2
Galactose	Galactose oxidase	0.01 – 1
Glucose	Glucose oxidase/catalase	0.001 – 0.8
Glucose	Hexokinase	0.5 – 25
Lactate	Lactate 2-monooxygenase	0.005 – 2
Lactate	Lactate oxidase/catalase	0.002 – 1
Lactose	β-Galactosidase + glucose oxidase/catalase	0.05 – 10
Oxalic acid	Oxalate oxidase	0.005 – 0.5
Oxalic acid	Oxalate decarboxylase	0.1 – 3
Penicillin	β-Lactamase	0.01 – 500
Sucrose	Invertase	0.05 – 100
Triglycerides	Lipoprotein lipase	0.1 – 5
Urea	Urease	0.01 – 500
Uric acid	Uricase	

If the solutions used, as well as the samples, are filtered through a 1–5 micron filter and if microbial growth in the solutions and the flow lines is prevented, good operational stability with unchanged performance for many samples (thousands) will be achieved. The column may then function for several months.

Normally the thermograms are evaluated by peak height determination, which is simple to perform both manually from the chart recorder output and automatically by an integrator or computer. The area under the peak and the ascending slope of the peak have also been found to be linearly related to the substrate concentration (16), but peak height determination tends to give higher precision (17).

5. Applications

Very satisfactory results have been obtained with thermal biosensors in many bioanalytical fields, including clinical chemistry, process control, fermentation monitoring and other analyses within biotechnology, and environmental control. *Table 2* summarizes a number of assays that have been studied with enzyme thermistors and similar devices over many years. This means that the detection ranges given in some cases could be improved with modern equipment. Most of the references to the original literature can be found in refs 2 and 3. Some of these assays will be described in some detail below.

5.1 Applications based on immobilized enzymes

Glucose

Of all metabolites glucose has been the subject of most studies using biosensors, including thermal analysers. This is because glucose determination is one of the most common clinical analyses and because glucose is a very common substrate in fermentation technology, two fields where dedicated analysers can be expected to find successful applications. Two enzymes have been employed for glucose determination with thermal biosensors: hexokinase (9) and glucose oxidase. The latter is usually co-immobilized with catalase (2,10) which more than doubles the total enthalpy change, eliminates the deleterious effects of the hydrogen peroxide formed in the glucose oxidase reaction, and restores half of the oxygen consumed by this reaction. The advantages of using glucose oxidase/catalase are high sensitivity through the large molar enthalpy change of the reactions, high specificity and no need for added cofactors. The useful concentration range is, however, limited to less than 0.7−1.0 mM by the supply of oxygen, which is soluble only to a small extent in water, about 0.25 mM at 25°C. The limit of detection can be as low as 1 μM.

By using a more soluble electron acceptor than oxygen, such as benzoquinone (11), the useful working range can be considerably increased, at least to 75 mM. The linear range for an assay based on hexokinase lies within 0.1−25 mM. With respect to sensitivity both assays are consequently very useful for clinical as well as for biotechnological glucose determinations. With glucose oxidase we find a precision (relative standard deviation) of 0.6% for samples determined within a day. A column can be used for several months and for several hundreds of determinations on serum samples (2). Glucose concentrations measured by the enzyme thermistor correlate well with the values obtained from conventional, photometric, enzymic techniques used in routine hospital diagnoses.

Disaccharides

The glucose oxidase/catalase-based assay can be used together with a disaccharide splitting enzyme for the determination of disaccharides containing glucose, such as cellobiose and lactose (2). Such hydrolytic enzymes are usually associated with too low an enthalpy change to be directly useful in thermometric assays, with the exception of invertase which can be used for the direct determination of sucrose in the range of 0.05−100 mM. Thus sucrose can be conveniently measured in biotechnological samples in the presence of glucose (18). Other examples of carbohydrate analyses are the determination of galactose with galactose oxidase (EC 1.1.3.9) and the determination of vitamin C with ascorbate oxidase (EC 1.10.3.3) with a linearity between 0.05 and 0.6 mM (19).

Urea

Another metabolite that has attracted much interest for biosensor studies is urea (6,14). With urease, which gives a linear range of at least 0.01−200 mM, a clinically useful assay can be designed with a precision better than 1%. Urease also results in a very stable enzyme column, provided that it is protected against heavy metals by the addition of 1 mM EDTA and 1 mM reduced glutathione for example.

Triglycerides

Practical useful methods for the determination of the main blood lipid classes have been developed (20). Triglycerides may be determined with lipoprotein lipase (EC 3.1.1.34) immobilized on CPG with a pore size of 2000 Å. A suitable assay buffer is 0.1 M Tris, pH 8.0, containing 0.5% Triton X-100. Using a split-flow apparatus, a linear response was obtained from 0.05 – 10 mM tributyrin and 0.1 – 5 mM triolein The triglyceride concentration in serum samples could be determined directly up to 3 mM after a 2-fold dilution with the Tris buffer. Good correlation with conventional spectrophotometric enzymic methods was obtained.

Phospholipids

Phospholipids can be determined with the use of three consecutively acting enzymes (20). Phospholipase D (36 IU) is added directly to a 0.05 ml sample, which is then injected into the buffer stream, 1 ml min^{-1} of 0.1 M Tris−HCl, pH 8.0, containing 15 mM $CaCl_2$ and 0.5% Triton X-100. The enzyme column is loaded with choline oxidase and catalase co-immobilized on CPG giving a linear range of 0.03 – 0.19 mM. This is sufficient for the direct determination of phospholipids in 10-fold diluted serum samples and gives a good correlation with conventional methods. By applying the rather unstable phospholipase in soluble form, the whole system performs well over 8 weeks and for at least 1600 analyses with each column.

Cholesterol and cholesterol esters

Cholesterol is dissolved in 0.16 M phosphate buffer, pH 6.5, containing 12% (v/v) ethanol and 8% (v/v) Triton X-100 and measured either with cholesterol oxidase immobilized on CPG in a differential ET unit or with a catalase-loaded ET using the cholesterol oxidase column as a pre-column. By treating the samples with cholesterol esterase, cholesterol esters may also be measured (20).

Triglycerides in organic solvents

We have recently started studies of thermometric measurements in organic solvents (21). One reason for this is that the lower heat capacity of organic solvents should lead to up to three times larger temperature change than in water at the same enthalpy change. Since the enzymes usually work well in organic solvents under certain conditions, it should also be interesting to try to develop assays for substrates which are difficult to dissolve in aqueous solutions. An example of this is the determination of triglycerides directly in cyclohexane, in which triglycerides are readily soluble. The enzyme, lipoprotein lipase, functions well in this solvent and the response is about 2.5 times higher than in water as shown in *Figure 3*.

L-Lactate

For L-lactate determination we have used two different enzymes with good results. With 50 units of lactate-2-monooxygenase (EC 1.13.12.4; from *Mycobacterium smegmatis*) applied to a CPG column, lactate can be determined with linear response in the range 0.005 – 1 mM in 0.2 M sodium phosphate buffer, pH 7.0. Alternatively, 25 U of lactate oxidase from *Pediococcus pseudomonas* (EC 1.1.3.2) is applied

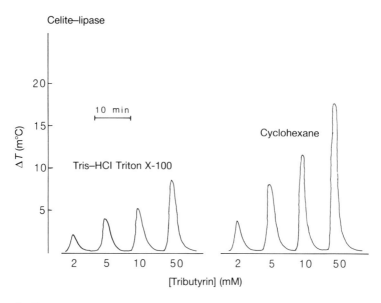

Figure 3. Response curves for 0.5 ml samples of glyceryl tributyrate obtained with a Celite – lipase loaded enzyme thermistor in 20 mM Tris – HCl, pH 8.0, containing 5 mM $CaCl_2$ and 4% Triton X-100 and in cyclohexane saturated with the same buffer. Flow rate: 1 ml min^{-1}. Working temperature: 30°C.

together with 100 000 U of catalase on a CPG column. The assay is run in 0.1 M sodium phosphate, pH 7.0, with a linear response in the range of 0.002 – 1 mM. Both enzyme preparations are stable for several months (3).

Oxalate

Two different enzymes have also been tried for the determination of oxalate. With oxalate decarboxylase (EC 4.1.1.2) a working range of 0.1 – 3 mM was obtained, but with rather poor enzyme stability. Much better enzyme stability is attained with an oxalate oxidase (EC 1.2.3.4) from barley seedlings (22), which is stable for 3 – 4 weeks and gives a useful analytical range of 0.005 – 0.5 mM in 0.1 M sodium citrate buffer, pH 3.5, containing 2 mM EDTA and 0.8 mM 8-hydroxyquinoline. The assay is suitable for urine samples, which should be diluted 10-fold and passed through a C_{18}-cartridge to remove interferences, as well as for beverage and food samples.

Ethanol

Ethanol can be measured with alcohol oxidase (EC 1.1.3.13) from *Candida boidinii*. To 1 g of CPG is added 50 U of the enzyme together with 130 000 U of catalase. Co-immobilization with catalase has a profound effect on the stability of the enzyme column, rendering it stable for several months with an operating range of 0.01 – 2 mM using 0.1 M sodium phosphate buffer, pH 7.0. This assay is useful for samples from beverages and for blood samples.

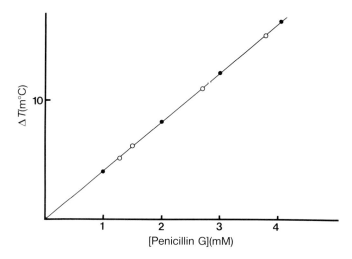

Figure 4. Calibration curve for Penicillin G in 0.3 M sodium-phosphate pH 7. (●), with samples containing known amounts of penicillin G in 10-fold diluted fermentation broth plotted in the same diagram (○). The ET column contained CPG-bound β-lactamase. Flow rate: 1 ml min^{-1}. Sample volume: 1 ml.

β-Lactams

Very successful assays have been designed for β-lactams, such as penicillin G and V (17) using β-lactamases, such as penicillin type I from *Bacillus cereus* (EC 3.5.2.6). The useful linear range is about 0.01 – 100 mM. The enzyme columns are very stable and useful for several months or for thousands of samples, with a very good correlation with other techniques, such as HPLC. Several enzyme thermistors are used in industrial laboratories for penicillin analysis. *Figure 4* shows a calibration curve for penicillin G together with some samples from fermentation broth.

Table 3 concludes this section by summarizing some important enzyme thermistor assays based on immobilized enzymes as well as by giving recommended working conditions.

5.2 Applications based on immobilized cells

In those cases where the pure enzyme is too unstable or difficult to isolate or in case the desired reaction involves sequential enzyme actions, it might be advantageous to use immobilized whole cells instead of enzymes (3). Coenzyme regeneration is, for instance, 'automatically' handled by the cell, so the buffer stream does not need to be supplemented with expensive coenzymes. Furthermore, immobilized cells can be used in assays based on the metabolic effect of an agent, such as a vitamin or a poison. A disadvantage is the lack of specificity in comparison with enzymes. The general applicability is a valuable feature of thermometric instruments, which does

Table 3. A summary of important enzyme thermistor assays.

Analyte	Enzyme column	Buffer	Useful range (mM)	Reference
Ethanol	50 U of alcohol oxidase (EC 1.1.3.13) + 130 000 U of catalase g^{-1} of CPG (1350 Å)	0.1 M sodium-phosphate, pH 7.0	0.01 – 2	3
Glucose	1000 U of glucose oxidase + 130 000 U catalase ml^{-1} of CPG (550 Å)	0.1 – 0.2 M sodium-phosphate, pH 7.0	0.002 – 0.8	3
Lactate	25 U lactate oxidase (EC 1.1.3.2) + 130 000 U catalase ml^{-1} CPG (1350 Å)	0.1 M sodium-phosphate pH 7.0	0.005 – 1	3
Oxalate	4 U oxalate oxidase ml^{-1} CPG (800 Å)	0.1 M sodium citrate, pH 3.5 + 2 mM EDTA + 0.8 mM 8-hydroxy-quinoline	0.005 – 0.5	22
Penicillin	1000 U β-lactamase ml^{-1} CPG (500 Å)	0.15 M sodium-phosphate, pH 7 + 1 mM NaN$_3$	0.01 – 100	17
Sucrose	10 000 U invertase g^{-1} of CPG (1350 Å)	0.1 M sodium-citrate, pH 4.6	0.1 – 100	19
Urea	600 U of urease ml^{-1} CPG or Eupergit C	0.1 M sodium-phosphate, pH 7 + 1 mM EDTA + 1 mM reduced glutathione	0.01 – 100	14

not limit one to follow just oxygen consumption or ammonia production, the common situation with electrode-based systems.

There are many suitable cell immobilization techniques at hand (23). It is important to maintain a sterile environment in order to let the desired reaction (sequence) be the dominating one and to make the beads sufficiently small, preferably less than 1–2 mm, to shorten response time. For example, *Gluconobacter oxydans* can be immobilized in a calcium alginate gel for different calorimetric studies (3). A cell suspension containing 0.2 g ml^{-1} (dry wt) of the bacterial cells is mixed in a 1% alginate solution in the proportions 1:4 (w/w) and extruded through a 1 mm orifice into a 0.1 M CaCl$_2$ solution. After standing for 2 h, the preparation is cut into 2 mm pieces and packed in a 1 ml ET column.

This column can be used for the determination of glycerol in 0.1 M sodium succinate, pH 5.0, containing 10 mM $CaCl_2$. At a flow rate of 0.9 ml min^{-1} the response curve is linear up to 2 mM for sample pulses with a sensitivity of 0.003° per mM. The response is slower than with an enzyme column, but allows for up to 15–20 samples per hour to be analysed.

In our opinion the use of immobilized cell columns in studies of metabolic effectors has great potential due to its general applicability. The metabolic heat is very likely to be affected by, for instance, the presence of a metabolic poison, such as a nerve gas, and thus an ET with a cell column can be used as a toxin monitor. Substances with an activating effect, such as vitamins, can also be detected (3). A disadvantage of this type of assay is that they often become one-shot sensors or possibly integrating sensors. If large amounts of the sensing cells are available, the assay can be designed in a continuous-flow fashion (see Section 5.5) with a continuous response.

5.3 Chemical and enzymic amplification

The total temperature signal can be increased by the addition of sequentially acting enzymes. As already mentioned this is usually done in connection with hydrolytic enzymes, such as disaccharide splitting enzymes, which can be combined with an enzyme acting on the monosaccharide formed. A typical combination is β-galactosidase followed by glucose oxidase/catalase for the determination of lactose. We have already mentioned that oxidases are almost always combined with catalase or occasionally with peroxidase for two reasons: besides more than doubling the total enthalpy change, the damaging effects of hydrogen peroxide are prevented.

In cases where a proton is formed (or taken up) by the enzymic reaction, as with proteolytic enzymes, a considerably greater total heat production can be obtained by using a buffer with a high protonation enthalpy, such as a Tris buffer (16).

We have found that amplification of the sensitivity by substrate or coenzyme recycling is especially efficient in thermometric analysis since in some cases all the reactions involved contribute to an increased total enthalpy change. For example, a more than 1000-fold enhanced sensitivity could be observed for lactate or pyruvate by using co-immobilized lactate oxidase and lactate dehydrogenase (LDH) (24). L-Lactate is oxidized by lactate oxidase to pyruvate, which is reduced to lactate again by LDH. The total enthalpy change of this system can be further increased by the addition of catalase to the enzyme column making the overall enthalpy change as high as -225 kJ mol^{-1}, see *Figure 5*. By this arrangement lactate (or pyruvate) concentrations as low as 10 nM can readily be determined.

A coenzyme recycling system for the determination of ATP/ADP has been recently studied (25) in which concentrations as low as 10 nM can be determined using pyruvate kinase and hexokinase. This cycle produces pyruvate, and an enormous amplification can be realized by combining this cycle with the pyruvate recycling system presented above. Another example of a coenzyme recylcing arrangement is presented in ref. 3. NAD^+ or NADH is recycled by the two enzymes lactate dehydrogenase and glucose-6-phosphate dehydrogenase, resulting in an up to 80-fold improvement in the sensitivity.

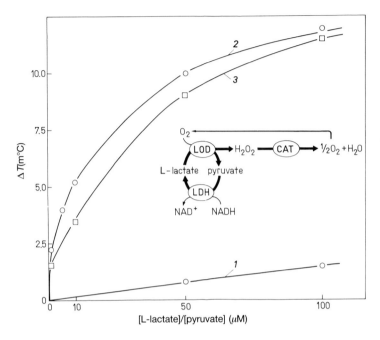

Figure 5. Temperature response for 0.5 ml samples of (**1**) L-lactate with only the lactate oxidase-catalase system working (NADH absent); (**2**) L-lactate with all enzymes. Lactate dehydrogenase and lactate oxidase catalase working (1 mM NADH added); (**3**) the same as in (2) with pyruvate as sample. The insert shows the action of the three enzymes co-immobilized in the ET column. (From ref. 24).

5.4 Enzyme activity determinations

The ET unit can be used for the determination of the activity of soluble enzymes, with a small modification of the flow system and with an empty or inactive column, preferably made of Teflon. The sample solution and buffer containing an appropriate substrate in excess, are passed through heat exchangers, and thoroughly mixed. The mixture is then passed through a short heat exchanger to eliminate mixing and solvation heats (26). Each flow channel has one long heat exchanger tube in contact with the outer aluminium jacket and one short tube in contact with the inner heat sink and can simply be adapted for this procedure. Even a residence time of less than 1 minute is enough for enzyme activity determinations with a sensitivity down to 0.01 U ml^{-1}. This technique has general applicability when the sensitivity is sufficient. It can be used with both sample pulses and with continuous sample introduction, and should be of particular interest for monitoring enzyme purification processes (26) as illustrated in *Figure 6*. The sensitivity is only about the same as for A_{280} monitoring, but it is of considerable importance to have a direct,

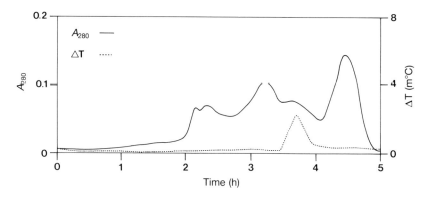

Figure 6. Gel filtration of 1 ml crude yeast extract on an Ultrogel AcA 44 column eluted with 0.2 M Tris−HCL, 0.0133 M $MgCl_2$, pH 7.8, at a flow rate of 0.75 ml min^{-1}. For on-line ET assay of hexokinase (dotted line) the effluent was mixed with substrate solution (0.54 M glucose and 0.011 M ATP) at a flow rate of 0.2 ml min^{-1}. (From ref. 26).

continuous-flow method that can be used on crude samples and with cheap substrates. We are currently investigating how this technique can be used for controlling downstream processes, for example coupled chromatographic procedures.

An alternative technique, which is of particular importance at low enzyme concentration, involves enrichment of the enzyme by affinity binding (preferably reversible) to an affinity column in the ET unit. The activity is determined by introducing a pulse of substrate in excess (3). This technique could also be used for the reversible immobilization of labile enzymes in cases when the enzyme column needs to be renewed frequently.

5.5 Measurements on cell suspensions

In a similar way the ET unit can be used for studies of the total metabolic heat from a cell suspension by mixing the cell stream with a substrate stream, or simply by letting cells, for instance from a fermentor, flow through an empty column in the ET unit. Aerobic cells must be run at a sufficiently low concentration in order not to deplete the available oxygen while the cells pass through the flow system. This technique should be useful in studies of metabolic effectors and in studies of toxic effects, and as a general method to follow the state of a fermentation. The heat or the temperature level recorded represents the total metabolism of microorganisms, and changes in the metabolism are directly detectable as changes in the temperature level. For example, in a study on the ampicillin susceptibility of *Escherichia coli* (27), changes in the metabolic behaviour during cultivation under standardized conditions were clearly observable, even in the presence of ampicillin at a concentration 10 times lower than the minimal inhibitory concentration (MIC).

5.6 Process monitoring and control

Most analytical techniques used today for process control are discontinuous, off-line procedures and on-line analyses are available only for simple measurements, such as pH and pO_2. Therefore we have developed a great interest in on-line applications of thermal biosensors. Because of space limitations only a few remarks will be made here. Determinations of various sugars as well as of ethanol and β-lactams as described in Section 5.1 are of immediate interest for process control. An evaluation was made of how well the degree of conversion by an invertase reactor could be controlled by the ET signals from a continuous determination of sucrose by invertase and from the determination of glucose (using glucose oxidase/catalase) by a flow injection technique repeated every 5 min (18). In these experiments, which lasted several hours, both types of monitoring worked very satisfactorily, but with experiments of longer duration, the baseline needed to be checked regularly during continuous measurement (about 5 times per day). Repeated flow injection analysis may then be the preferred technique, and can easily be arranged with the aid of an injection valve, controlled by a simple timer or by a computer. Samples from a fermenter, and other crude samples, must of course be cleaned up before they can be introduced on to a immobilized enzyme column (28). This can be made on-line with a dialysis unit or with a tangential flow ultrafiltration or sterile filtration unit. For use in fermenters under sterile conditions, we have designed an autoclavable dialysis/filtration probe that is inserted through the fermenter wall as described in ref. 29, which also gives some more examples of process control by enzyme thermistors.

It should once again be noted that the capability to register the power−time curve (thermogram) of a fermentation process as an indication of the total metabolism (as described in the former section) also provides valuable information for process control.

5.7 Environmental control applications

We have exploited the calorimetric detection principle in two different ways in environmental control analysis. The heat of the conversion of a pollutant by an enzyme, or by a cell metabolic route, may in some cases be sufficient for measurement, at least in process streams or in sewages. Thus, the enzyme rhodanese (EC 2.8.1.1) can be used for the determination of cyanide, and there are enzymes available that can use pesticides as substrates (16). The other alternative, which is much more sensitive, is to measure the inhibitory effect of a pollutant (3). To detect an environmental poison through its biological effect is actually an ideal concept. As an example, we have developed a technique for a highly sensitive determination of heavy metals (Hg^{2+}, Cu^{2+} and Ag^+) based on the inhibition of urease. A relatively small amount of urease (1000 U) is applied to 1 ml of CPG. The response obtained for a 0.5 ml pulse of 0.5 M urea in phosphate buffer at a flow rate of 1 ml min^{-1} is noted, whereafter 0.5 ml of the sample is injected. Thirty seconds after the sample pulse (timing is important), a new 0.5 ml pulse of 0.5 M urea is injected

Table 4. Application fields of thermal biosensors.

Mode	Column	Sample type
Metabolite determination	Immobilized enzymes or cells	Physiological fluids; Fermentation broths
Inhibition; toxic monitoring	Immobilized enzymes or cells	Environmental control and water samples
Enzymic activity	Inactive. Mix with substrate solution	Enzyme solutions, e.g. at enzyme purification
Cell metabolism	Empty. Mix with substrate solution	Microbial suspensions
TELISA	Immunosorbent	Proteins, hormones

and the response compared with the non-inhibited peak. After a sample has been run, the initial response can be restored by washing the column with 0.1–0.3 M NaI plus 50 mM EDTA for 3 min. A calibration curve can be made with standard samples and the heavy metal concentration of an unknown sample can be estimated from such a curve. For the conditions given here, a 50% inhibition (50% of the initial response) could be expected for a 0.5 ml pulse of 0.04–0.05 mM Hg^{2+} or Ag^+ or 0.3 mM Cu^{2+}. Longer sample pulses result in a considerably higher sensitivity.

5.8 TELISA (Thermometric enzyme linked immunosorbent assay)

For rapid and sensitive determination of larger molecules, such as hormones and antibodies which are present in, for instance, fermentation broths, an automated, flow through thermometric ELISA has been designed. The use of this assay for the determination and monitoring of the production and release of human proinsulin by genetically engineered *E.coli* cells, has recently been reported (30). Anti-insulin serum was affinity purified against beef insulin. The fraction which elutes with 0.2 M glycine–HCl, pH 2.2, is coupled, after dialysis against 0.2 M $NaHCO_3$, to Sepharose 4B (0.2 mg g^{-1} wet gel) activated with tresyl chloride (6 μl g^{-1} wet gel). A beef insulin–peroxidase conjugate is prepared (30) and the fraction with a weight ratio of insulin to peroxidase of 1:2 is isolated by chromatography on Sephacryl S-200.

The unlabelled antigen in the sample (or standard) is mixed with a fixed amount of enzyme labelled antigen and the mixture is then applied to the immunosorbent column mounted in the ET. By increasing the amount of unlabelled antigen in the sample, the amount of enzyme-labelled antigen bound to the column decreases. The amount of peroxidase-labelled antigen bound to the column is then determined by measuring the enzymic activity thermometrically when a substrate pulse (2 mM

H$_2$O$_2$, 14 mM phenol and 0.8 mM 4-aminoantipyrine) is injected. Finally, bound antigen is removed from the immunosorbent by washing with 0.2 M glycine−HCl, pH 2.2, thereby regenerating the column for the next assay. The response time of the automated assay, which can be controlled by a programmable controller (Hizac D28, Hitachi) operating valves for different eluants and substrates and the sample changer, is 7 min after sample introduction, and a single assay cycle is completed after 13 min. In this study insulin concentrations in the range of $0.1-50$ μg ml^{-1} were determined with a good correlation with conventional radioimmunoassay. Standard curves are reproducible over a period of several days, even when the immunosorbent column is stored inside the ET. Consequently, this method is very suitable with respect to time, sensitivity and stability for monitoring fermentations. It is also useful in other situations where rapid determinations on a limited number of samples are desired, for instance, to establish a proper dosage of drugs in medicine.

6. Concluding remarks

Due to their general detection principle, thermal analysers have a multitude of application possibilities as indicated by *Table 4* which summarizes the application fields we have studied. In clinical chemistry there is an increasing interest in biosensors for their use in decentralized health care. Biosensors should also be very suitable for home monitoring of certain metabolites, such as glucose in diabetes control. In the case of home monitoring, thermal biosensors should be of special interest because of their easy operation and maintenance and their general applicability. In the rapidly growing and important field of biotechnology there will be an increasing demand for flow stream analysers in process control, fermentation monitoring, and downstream analyses. Thermal flow analysers appear to be particularly attractive for downstream analyses due to the possibility of analysing turbid, particulate, or coloured samples using highly specific continuous-flow techniques, not only for enzyme substrates, but also for enzymes, cells (metabolic activity), and proteins (TELISA). In environmental control the possibility for using very sensitive assays directly related to the biological effect of the pollutant should be of special interest.

References

1. Spink, C. and Wadsö, I. (1976). *Methods Biochem. Anal.,* **23**, 1.
2. Danielsson, B. and Mosbach, K. (1986). In *Biosensors: Fundamentals and Applications.* Turner, A. P. F., Karube, I., and Wilson, G. (eds), Oxford University Press, Oxford, p. 575.
3. Danielsson, B. and Mosbach, K. (1988). In *Methods in Enzymology.* Mosbach, K. (ed.), Academic Press Inc., London and New York, Vol. 137, p. 181.
4. Mosbach, K. and Danielsson, B. (1974). *Biochim. Biophys. Acta,* **364**, 140.
5. Weaver, J. C., Cooney, C. L., Fulton, S. P., Schuler, D., and Tannenbaum, S. R. (1976). *Biochim. Biophys. Acta,* **452**, 285.

6. Tran-Minh, C. and Vallin, D. (1978). *Anal. Chem.*, **50**, 1874.
7. Rich, S., Ianiello, R. M., and Jespersen, R. D. (1979). *Anal. Chem.*, **51**, 204.
8. Pennington, S. N. (1976). *Anal. Biochem.*, **72**, 230.
9. Bowers, L. D. and Carr, P. W. (1976). *Clin. Chem.*, **22**, 1427.
10. Schmidt, H. -L., Krisam, G., and Grenner, G. (1976). *Biochim. Biophys. Acta*, **429**, 283.
11. Kiba, N., Tomiyasu, T., and Furusawa, M. (1984). *Talanta*, **31**, 131.
12. Maramatsu, H., Dicks, J. M., and Karube, I. (1987). *Anal. Chim. Acta*, **197**, 347.
13. Muehlbauer, M. J., Guilbeau, E. J., and Towe, B. C. (1989). *Anal. Chem.*, **61**, 77.
14. Danielsson, B., Gadd, K., Mattaisson, B., and Mosbach, K. (1976). *Anal. Lett.*, **9**, 987.
15. Mattiasson, B., Danielsson, B., and Mosbach, K. (1976). *Anal. Lett.*, **9**, 867.
16. Danielsson, B., Mattaisson, B., and Mosbach, K. (1981). *Appl. Biochem. Bioeng.*, **3**, 97.
17. Decristoforo, G. and Danielsson, B. (1984). *Anal. Chem.*, **56**, 263.
18. Mandenius, C. F., Bülow, L., Danielsson, B., and Mosbach, K. (1985). *Appl. Microbiol. Biotechnol.*, **21**, 135.
19. Mattiasson, B. and Danielsson, B. (1982). *Carbohydr. Res.*, **102**, 273.
20. Satoh, I. (1988). In *Methods in Enzymology*. Mosbach, K. (ed.), Academic Press Inc, London and New York, Vol. 137, p. 217.
21. Flygare, L. and Daneilsson, B. (1989). *Ann. NY Acad. Sci.*, **542**, 485.
22. Winquist, F., Danielsson, B., Malpote, J. -Y., Persson, L., and Larsson, M. -B. (1985). *Anal. Lett.*, **18**, 573.
23. Nilsson, K., Scheier, W., Katinger, H. W. D., and Mosbach, K. (1987). In *Methods in Enzymology*. Mosbach, K. (ed.), Academic Press Inc, London and New York, Vol. 135, p. 399.
24. Scheller, F., Siegbahn, N., Danielsson, B., and Mosbach, K. (1985). *Anal. Chem.*, **57**, 1740.
25. Kirstein, D., Danielsson, B., Scheller, F., and Mosbach, K. (1989). *Biosensors*, **4**, 231.
26. Danielsson, B., Bülow, L., Lowe, C. R., Satoh, I., and Mosbach, K. (1981). *Anal. Biochem.*, **117**, 84.
27. Hörnsten, E. G., Danielsson, B., Elwing, H., and Lundström, I. (1986). *Appl. Microbiol. Biotechnol.*, **24**, 117.
28. Wehnert, G., Sauerbrei, A., Bayer, Th., Scheper, Th., Schügerl, K., and Herold, Th. (1987). *Anal. Chim. Acta*, **200**, 73.
29. Mandenius, C. F. and Danielsson, B. (1988). In *Methods in Enzymology*. Mosbach, K. (ed.), Academic Press Inc, London and New York, Vol. 137, p. 307.
30. Birnbaum, S., Bülow, L., Hardy, K., Danielsson, B., and Mosbach, K. (1986). *Anal. Biochem.*, **158**, 12.

9

Theoretical methods for analysing biosensor performance

MARK J. EDDOWES

1. Introduction

1.1 Role of theory

Theory, along with experiment, forms a vital part of a practical scientific approach. Though perhaps redundant on its own, theory enables the most efficient use of time spent performing experiments. This is important since experiments are often difficult and time consuming and will not always be successful. Proper use of theory assists effective planning of experiments and enables the researcher to gain the maximum amount of information from each experiment performed. In applied work, such as biosensor research and development, theory used in a predictive manner can identify which, of perhaps a multitude of approaches, is most likely to yield the desired result. It is therefore important that proper consideration be given to theory and its application to the solution of practical problems involving real systems. The aim of this chapter is to demonstrate how, in practice, the problem of the provision of a theoretical description of biosensors may be approached and illustrate this approach by way of some examples commonly encountered.

1.2 Fundamental aspects of biosensor operation

Though a wide variety and ever increasing number of sensing methods have been employed in biosensors, some fundamental principles are generally common to all. First, we may note that biological components, either enzymes, antibodies or receptors, are employed to provide specificity. This specificity derives from biological recognition based on the binding of the target analyte by the biological component employed and consequently a reaction sequence involving such a binding step will generally be encountered. Second, sensing will generally take place at a surface and transport to this surface prior to reaction at it will frequently be an important factor. Third, pseudo-steady-state conditions, in which the sensor signal is dependent upon a reaction proceeding at a constant rate, will often be encountered, for example in enzyme-based and amperometric systems. The kinetics of transport coupled with surface reaction(s) will therefore be important in such cases. Even where a true equilibrium measurement is to be made, the rate of approach to equilibrium will be important in determining the response time. In this case the kinetics of the process will again require consideration. These fundamental principles will first be reviewed before proceeding to consider how the theoretical description of a real biosensor system, based upon them, can be derived.

1.3 Mathematical tools

Provision of a formal theoretical description relies upon the available mathematical tools. Since, in considering biosensors, a chemical change will generally be involved, these will most commonly be based upon differential calculus, the mathematics of change. Further, with parameters changing both with time and with distance from the surface at which the reaction takes place, problems in partial differential calculus will often be encountered. The purpose of the solution of these differential equation systems will be to provide a description of the behaviour of the system in a more manageable, frequently algebraic form, from which predictive studies or experimental data analysis can be readily performed. For the non-specialist in mathematics this need not be as daunting as it might at first appear. In a great many cases solutions to the more complex parts of the problem are known. Theoretical analysis will then involve the relatively straightforward process of linking together the different components of the system rather than starting completely afresh. Often, solutions to analogous problems in other scientific disciplines, for example chemical engineering and heat transport, can be directly employed. Even where solutions differ, they can show, by way of examples, the approach required in tackling a similar problem. The non-specialist in mathematics is therefore not alone when addressing a new problem but has a wealth of examples on which to draw. Some of the mathematical tools available for the solution of the types of problems encountered in analysis of biosensors are described below and their application is illustrated.

2. Biological reactions

The biological component of a biosensor is typically employed to confer specificity upon the sensing system and to provide recognition for analytes which are otherwise difficult to detect selectively. The specific recognition characteristics of enzymes, antibodies and general receptor systems are employed to perform a variety of functions within the organisms from which they are isolated; catalysis, defence, communication and control being the primary ones. This specific recognition is based upon chemical binding of the one component to its complementary partner, these being the target analyte and the biological component of the biosensor respectively. This chemical binding can be very strong indeed and the binding process is frequently very rapid, often close to the limit set by the diffusional encounter in the case of antibody−antigen binding, for example. Whatever the normal function of a biological component within an organism, the basic principle behind its capacity for specific recognition will be the same and, for our current purposes, we need concern ourselves only with the fundamentals of the binding process as it relates to biosensors.

2.1 Homogeneous single site binding equilibrium

We can consider, as an example of a simple binding equilibrium, the reaction (1–3) between an antibody, (Ab), and its antigen, Ag, to form the complex, Ab−Ag. The general principles and relationships derived will be generally applicable to any simple adduct formation reaction. The association reaction will be characterized by the bimolecular, or second-order rate constant, k_a, and the dissociation reaction by the first-order rate constant, k_d. The corresponding association and dissociation rates, v_a and v_d, will be given by:

$$v_a = k_a[\text{Ab}][\text{Ag}] \tag{1a}$$

$$v_d = k_d[\text{Ab}-\text{Ag}] \tag{1b}$$

where the square parentheses indicate concentration of the given chemical species. At equilibrium, the concentrations of reactants, Ab and Ag, and product Ab−Ag, will be such that the rates of association and dissociation are equal. Equating 1a with 1b and rearranging gives an expression for the equilibrium constant, K_a, for association, according to:

$$K_a = \frac{[\text{Ab}-\text{Ag}]}{[\text{Ab}][\text{Ag}]} = \frac{k_a}{k_d} \tag{2}$$

where the units of K_a are M^{-1}. Note that the association equilibrium constant has been defined in equation 2, the concentration of the product complex, Ab−Ag, being the numerator and the product of the concentrations of the reactants, Ab and Ag, the denominator in equation 2. In certain cases it is more convenient to work in terms of the dissociation constant, K_d, which is simply the reciprocal of the association constant, $1/K_a$, in units of concentration, M, where the forward reaction is considered to be dissociation of the reactant complex, Ab−Ag, to form dissociation products, Ab and Ag.

Frequently, when performing experiments, or indeed when making chemical measurements with a biosensor, the starting concentration of one component, the biological recognition component, Ab, for example, will essentially be fixed and the concentration of the other, in this case Ag, will be a variable or unknown. Under such circumstances the sum of concentrations of free antibody, $[\text{Ab}]_{eq}$, and bound antibody, $[\text{Ab}-\text{Ag}]_{eq}$, at equilibrium will be constant, independent of the added concentration of antigen. The relative proportion of free and bound antibody will vary with the equilibrium concentration of antigen, $[\text{Ag}]_{eq}$, and the nature of this variation will be of interest. Given a starting concentration of antibody, $[\text{Ab}]_0$, the equilibrium concentration, $[\text{Ab}]_{eq}$, will be simply $[\text{Ab}]_0 - [\text{Ab}-\text{Ag}]_{eq}$, the starting concentration less the concentration bound. Substituting this into equation 2 gives,

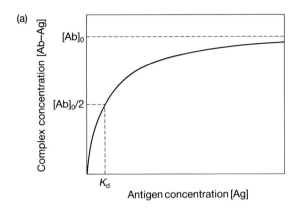

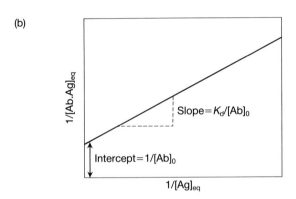

Figure 1. (a) Variation of complex concentration, [Ab − Ag], with concentration of binding species, [Ag]; (b) double reciprocal plot of the same data as in (a).

after rearrangement, an expression for the concentration of complex as a function of antigen concentration:

$$[Ab-Ag]_{eq} = \left(\frac{[Ag]_{eq}}{K_d + [Ag]_{eq}}\right)[Ab]_0 \qquad (3a)$$

which is often more conveniently expressed in a reciprocal form, according to:

$$\frac{1}{[Ab-Ag]_{eq}} = \frac{1}{[Ab]_0} + \left(\frac{K_d}{[Ab]_0}\right)\frac{1}{[Ag]_{eq}} \qquad (3b)$$

The form of this relationship is illustrated in *Figure 1a* and *b*. The double reciprocal plot of $1/[Ab-Ag]_{eq}$ against $1/[Ag]_{eq}$ is linear and the convenience of its use, particularly with respect to determination of the dissociation constant from experiment data, can be seen from *Figure 1b*. The plot provides two pieces of information about the system. The first, a value for $1/[Ab]_0$, is obtained from the intercept of the plot on the $1/[Ab-Ag]_{eq}$ axis and the second, a value for $K_d/[Ab]_0$ and hence K_d, is obtained from the slope. Effectively the same analysis can be achieved by the Scatchard plot (1) of $[Ab-Ag]/[Ag]$ against $[Ab-Ag]$, which is perhaps the most widely used method for obtaining affinity constant and binding site concentration values from experimental data. The straightforward double reciprocal plot tends to concentrate, into one region of the graph, data taken at equal steps in antigen concentration and it is for this reason that the Scatchard plot, which spreads the data more evenly, is often preferred.

It should be noted that the concentration of antigen at equilibrium, $[Ag]_{eq}$, in equation 3 does not represent the total amount of antigen in solution. No account has been made of the amount of antigen bound in the complex in the way that this has been taken into account for the antibody. However, where the one component, in this case Ag, is in sufficient excess such that $[Ab-Ag]_{eq} < < [Ag]_{eq}$, there is negligible difference between the total amount of antigen and the amount free at equilibrium. Under these circumstances this difference can therefore be neglected. Thus, equation 3 gives a general description of the binding equilibrium where the concentration of the one component is in considerable excess over the other. Where the initial concentrations of both components of the complex are of similar magnitude the simplification will not apply. An exact analytical expression for the concentration of the complex can be found by substitution of $[Ag]_{eq} = [Ag]_0 - [Ab-Ag]_{eq}$ into equation 2, as well as the analogous expression for $[Ab]_{eq}$. In this case a quadratic equation results which can be solved to give the exact expression for the complex in terms of the starting concentrations of both antibody and antigen. The general form of the complex versus antigen concentration plot is the same as that for the simplified case but significantly depressed. However, the double reciprocal plot can be very distinctly curved and extraction of the binding parameters from it is not straightforward.

2.2 Binding at surfaces: the adsorption isotherm

Binding at surfaces may often be of importance in biosensor analysis and this represents a specific case of the more general equilibrium described by equation 3. If we consider a surface with a certain density of binding sites, γ_{lim}, in units of moles m^{-2}, this can be equated directly with the initial concentration of antibody, $[Ab]_0$, in the description of complex formation in homogeneous solution. The whole of this earlier treatment can be directly applied to give an expression for the equilibrium surface coverage or surface concentration of the Ab$-$Ag complex, γ_{eq}:

$$\gamma_{eq} = \left(\frac{[Ag]_{eq}}{K_d + [Ag]_{eq}} \right) \gamma_{lim} \quad (4a)$$

$$\frac{1}{\gamma_{eq}} = \frac{1}{\gamma_{lim}} + \left(\frac{K_d}{\gamma_{lim}} \right) \left(\frac{1}{[Ag]_{eq}} \right) \quad (4b)$$

Alternatively, the surface equilibrium can be expressed in terms of the fractional surface coverage, $\theta_{eq} = \gamma_{eq}/\gamma_{lim}$, which is simply the equilibrium coverage normalized with respect to the limiting coverage. Equation 4 describe the Langmuir adsorption isotherm (4) which displays the characteristics of the binding equilibrium, illustrated in *Figures 1a* and *b*. As before, a double reciprocal plot of experimental data gives the binding parameters, in this case γ_{lim} and K_d, from the intercept and slope. It should be noted that, for the case given by equation 4, the solution species, Ag, is in considerable excess over the surface binding sites. The same general considerations as discussed for homogeneous solution binding apply (5) where Ag is not in sufficient excess.

2.3 Heterogeneity of binding sites, multivalence and cooperativity

The preceding treatment has assumed the simplest case, that of a single type of monovalent ligand. The complexity of the problem increases significantly when mixtures of binding sites with different affinities are considered. This is particularly important when dealing both with binding at surfaces, where heterogeneity is not uncommon, and with antibodies obtained from immune serum, since these will be polyclonal in nature and consequently display a range of affinities for the antigen. Where such heterogeneity in the affinity of binding sites occurs, antigen will be distributed between the available antibody sites according to the different affinities. The subsequent superposition upon one another of the various equilibria having different characteristic affinity constants results in the linearity of the simple double reciprocal or Scatchard plot being lost. Deviations from simple behaviour will also occur when multivalent ligands are involved and this clearly can be of importance to antibody binding given the bivalence of the IgG system. A further possible complication arises from cooperativity in multivalent or surface binding, where the binding of a ligand at one site affects the affinity of an adjacent site. Exact treatment of these more complex cases is difficult, though general accounts of the effects of the factors involved upon the observed binding behaviour of antibodies have been given (1–3). The importance of heterogeneity and lateral interactions has also been considered (4) for surface adsorption in general where alternative adsorption isotherms to that of the simple Langmuir model have been proposed. These more detailed aspects are worth bearing in mind, though their quantitative inclusion in the description of a biosensor system will often be precluded due to the complexity that they would introduce.

2.4 Enzyme kinetics

Binding of substrate to an enzyme appears to be a prerequisite for effective enzyme catalysis and the theoretical basis for this is well understood (6). The presence of this binding step determines the nature of the observed kinetics (7,8) for enzyme catalysis and this can be illustrated by considering the simplest enzyme reaction scheme. The scheme involves a single substrate, S, which combines with enzyme,

E, to give the intermediate enzyme–substrate complex, ES. This complex then undergoes reaction to produce the product, P, the overall scheme being summarized by:

$$E + S \underset{k_{-1}}{\overset{k_1}{\rightleftharpoons}} ES \overset{k_2}{\rightarrow} E + P \tag{5}$$

where k_1 and k_{-1} are the forward and backward rate constants for complex formation and k_2 is the rate constant for complex decomposition into product. From this reaction scheme, rates of formation and disappearance of the ES complex can be written:

$$v_f = k_1[E][S] \tag{6a}$$

$$v_d = k_{-1}[ES] - k_2[ES] \tag{6b}$$

where v_f is the rate of the bimolecular ES complex formation reaction and v_d is the rate of its disappearance, this latter rate being the sum of the rates of dissociation back to reactants and decomposition into product. Equation 8 can be solved by employing the steady-state assumption; that is, by assuming that the rates of formation and breakdown of the complex are equal. This steady-state assumption is central to many kinetic treatments, some of which are illustrated in Section 4. It is employed, almost without a second thought in many instances, and it is worthy of some further consideration. Immediately after the reactants E and S are first mixed the steady state will not apply; the concentration of the complex, ES, will be zero and the formation rate will greatly exceed that for breakdown. As the reaction proceeds, the concentration of ES will build up and the rate of breakdown of the complex will increase towards the rate of its formation. For the enzyme catalysed reaction of equation 5, there will typically be a considerable excess of substrate over enzyme, the latter being constantly regenerated during the reaction. In this case, the rates of formation and breakdown of the enzyme–substrate complex will effectively be able to reach a steady-state value, being equal to one another, before an appreciable amount of the substrate has been consumed. Using the steady-state assumption, which is valid almost immediately after the reaction has begun, a rate equation for the intermediate can be written:

$$\frac{d[ES]}{dt} = k_1[E][S] - k_{-1}[ES] - k_2[ES] = 0 \tag{7}$$

and, since the total concentration of enzyme, $[E]_0$, at all times will be the sum of concentrations in free and complexed forms $[E]+[ES]$, substituting $[E]=[E]_0-[ES]$, equation 7 becomes:

$$\frac{d[ES]}{dt} = k_1[E]_0[S] - (k_1[S]+k_{-1}+k_2)[ES] = 0 \tag{8}$$

Theoretical methods for analysing performance

Equation 8 can be solved for [ES] to give:

$$[ES] = \frac{[E]_0[S]}{K_M + [S]} \tag{9}$$

where K_M, the Michaelis constant, is a term in rate constants defined by:

$$K_M = \frac{(k_{-1} + k_2)}{k_1} \tag{10}$$

Substituting this into the rate expression for decomposition of the complex to form products gives the rate, v, of product formation:

$$v = k_2[ES] = \frac{k_2[E]_0[S]}{K_M + [S]} \tag{11}$$

Equation 11 is of exactly the same form as the expression for complex formation, described earlier by equation 3a. That is, the substrate concentration appears in the numerator and a constant plus the substrate concentration appear in the denominator. The variation of observed rate with substrate concentration therefore follows the same form as the variation of antibody−antigen complex concentration with antigen concentration shown schematically in *Figure 1a*. There are two important points to note from the rate expression. First, at lower substrate concentrations where $[S] << K_M$, the rate is proportional to substrate concentration and is inversely proportional to the Michaelis constant. Unlike K_d in the analogous binding equation 3, K_M is not a true equilibrium constant but is the ratio of rate constants for complex formation and disappearance, the latter comprising both simple dissociation back to reactants plus decomposition into products. Nevertheless, to some extent the Michaelis constant typically does reflect the binding affinity of the enzyme for its substrate. Second, at higher substrate concentration the rate reaches its maximum velocity, $k_2[E]_0$, limited by the amount of enzyme. So, with the two step reaction sequence of equation 5, we have two processes which can control the overall observed rate. According to the conditions, either the association reaction to form the enzyme−substrate complex or the catalysed reaction step within the complex which results in product formation, might represent the rate determining step, that is the slowest step that controls the overall rate of reaction.

Since the enzyme catalysis rate equation, equation 11, is of a form identical to that for complex formation, equation 3a, data analysis can again be performed (8) by rearrangement into a double reciprocal form:

$$\frac{1}{v} = \frac{K_M}{k_2[E]_0[S]} + \frac{1}{k_2[E]_0} \tag{12}$$

The Lineweaver−Burke plot of reciprocal rate against reciprocal substrate concentration is completely equivalent to the double reciprocal plot used in the determination of binding parameters, illustrated by *Figure 1b*. In this double reciprocal form the two possible rate determining steps have been separated, the first term on the right hand side of equation 12 representing the association process linked with the enzyme catalysed step and the second representing the enzyme catalysed step alone. The plot enables the determination of the rate of the enzyme catalysed step, $k_2[E]_0$, from its intercept and the Michaelis constant, K_M, reflecting the association step, from its slope. Such a double reciprocal representation can often be used to good effect in kinetic analysis of multistep reaction sequences to separate out the different possible rate determining steps. Its use will be illustrated, subsequently, in application to more complex reaction sequences. As with the determination of binding parameters from plots of experimental data, alternatives (7,8) to the simple double reciprocal are often preferred to prevent the clustering of too many data points into one region of the graph.

It is to be stressed that the preceding Michaelis–Menten kinetic treatment is the simplest available for enzyme catalysis and will not apply under certain circumstances. For example, even for the simple reaction scheme given in equation 5, the initial stages of reaction are not described by this treatment. More complex reaction schemes exist, involving, for example, the reverse reaction in which complex formation between the enzyme and product must be taken into account, multiple substrates with additional reaction steps and inhibition. The Michaelis–Menten scheme serves as an adequate example for the current purposes and fuller accounts of the reaction sequences encountered in enzyme catalysis and the methods for their analysis can be found elsewhere (7,8).

3. Mass transport

Where reaction takes place in homogeneous solution it proceeds at the same rate uniformly throughout the medium and it is therefore necessary to consider only the way in which concentrations of the various components change with time. On the other hand, if reaction takes place at a surface the concentrations of reactants and products will change locally at that surface. This leads to what is commonly called concentration polarization, that is, substantial differences in concentration over distances that are very significant on a molecular scale. Transport from areas of high concentration to areas of low concentration may then become an important factor in the overall reaction sequence and in the control of its rate. Three mechanisms of mass transport can occur in solution: diffusion, convection and migration, of which the first two will normally be the most important. The principles and mathematics of mass transport are described in detail in a number of texts (9–11) and relevant aspects are reviewed below.

3.1 Diffusion as a random process

Diffusion is dependent upon the random motion of species. Where concentration differences exist in a solution, this random motion will lead, in time and in the absence of any perturbation, to a random distribution of the species and so to a homogeneous solution of uniform concentration. This is a familiar process and can be understood as follows. If we consider molecules within a solution experiencing Brownian motion as they undergo collision with the solvent we can see that the probability of any given molecule moving in any given direction is the same as the probability of any other molecule moving in any other direction. Given this equality of probability for each molecule, the net probability of overall motion of molecules from one given volume element in solution to an adjacent one is dependent upon the difference in the number of molecules within each volume element of solution. This results in the net motion out of the element which contains the greater number of molecules. No net motion will occur where the number of molecules is the same in each element. That is to say, chemical species move from regions of high to low concentration, at a rate that is dependent upon the difference in concentration between the two regions, reaching a stable state when no concentration differences exist. Conceptually then, diffusive mass transport is quite a simple process.

3.2 Fick's laws of diffusion

In order to use this understanding of molecular diffusion as a phenomenon dependent upon random motion in our description of mass transport we need first to write it down in a formal mathematical manner. Since the process is concerned with change this will take the form of a differential equation. First, let us consider an arbitrary plane surface within a solution, as illustrated in *Figure 2a*. Let the solution on one side contain a concentration c of a chemical species and that on the other a concentration $c + \delta c$. With equal probability of movement of any given molecule in any given direction, the amount of material crossing the plane surface in either direction will simply be proportional to the concentrations on the two sides of it. Hence the net flux, j, of material across the surface in a given time will be directly related to the concentration difference, δc, or more precisely to the concentration gradient, $\delta c/\delta x$. This result, a more complete derivation of which is given elsewhere (9), is Fick's first law of diffusion in one dimension:

$$j = -D \left(\frac{\delta c}{\delta x} \right) \tag{13}$$

where the constant of proportionality, D, is the diffusion coefficient in units of $m^2 sec^{-1}$, and the sign is that appropriate for defining the flux towards the origin, $x = 0$, of the coordinate system.

Further similar considerations will lead us to a description of the way in which the flux of material from one small volume element to another leads to concentration changes within those volume elements. We consider a volume element of thickness δx and are concerned with the net flux in and out of it from the volume elements on either side, as illustrated in *Figure 2b*. The change in the amount of material in the middle element is simply the difference in net fluxes, δj, into it from the

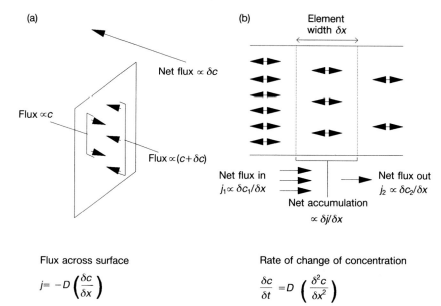

Figure 2. Schematic representation of (a) diffusional flux across a surface, and (b) the concentration change associated with this diffusional flux in the presence of a concentration gradient.

elements on either side of it, multiplied by the time, δt, and the area, A, over which the change is observed. Recalling from equation 13 that δj is $-D\delta(\delta c/\delta x)$ and that the change in amount is simply the concentration change, δc, multiplied by the volume, $A\delta x$, we can write $\delta c A\delta x = D\delta(\delta c/\delta x)A\delta t$, which describes the concentration change with time and distance. This, when rearranged into the appropriate differential form, is Fick's second law of diffusion in one dimension:

$$\frac{\partial c}{\partial t} = D\left(\frac{\partial^2 c}{\partial x^2}\right) \tag{14}$$

Fick's laws in one dimension have been presented for the sake of simplicity. Their forms in more dimensions and in a variety of coordinate systems are to be found in standard texts (10).

3.3 Application of Fick's laws to diffusion problems

As we stated earlier, since diffusive mass transport is a process of change, a consideration of it has produced a mathematical description in terms of differential equations. However, we will most frequently be concerned not just with the process of change of a system but with its state at a particular moment in time. That is, we typically wish to know concentrations of the chemical species involved in our sensing device at particular points in space at the time we make our measurement. In particular, we will be concerned with the rates of transport, i.e. the flux as described by Fick's first law, of material to the sensor surface and the build up or depletion of material there, in accordance with Fick's second law. In order to provide the required description in these terms we need to solve the differential equations of Fick's laws and to do this will require further information. This information, usually called the initial and boundary conditions, is a definition of the initial state of the system and its state at the boundaries of the region of space over which the diffusive mass transport process of interest is occurring. These boundary conditions will be dependent upon factors such as the biochemical and other processes occurring at the sensor surface and the concentration of species in the bulk of solution from which material is transported to it as well as geometrical aspects of the sensor's design.

We may consider, as an example of the application of Fick's laws, the diffusion of material to a planar surface at which it undergoes rapid reaction. Initially, before reaction has started, the concentration will be uniform throughout at its bulk solution value, c_b, which constitutes the initial condition. As regards the boundary conditions, with sufficiently rapid reaction at the surface, the surface concentration, c_0, will reduce effectively to zero and so a local concentration gradient is generated, providing the impetus for diffusional mass transport. The bulk concentration far from the surface, c_∞, will remain constant at its original starting value. The overall rate of such a reaction will be determined by the rate of mass transport to the surface and reaction is then said to be diffusion controlled. In order to obtain information about the state of the system and the reaction rate, equation 14 can be solved, subject to these initial and boundary conditions, by the Laplace transform method, as illustrated in standard electrochemical texts (9), to give an expression for the concentration as a function of both time and distance from the surface:

$$c = c_\infty \, erf\left(\frac{x}{2D^{1/2}t^{1/2}}\right) \qquad (15)$$

where erf is the error function. This expression describes the increase in concentration with increasing distance, x, away from the surface and the decrease in concentration with increasing time, t, and can be used to construct concentration profiles at different moments in time, as illustrated in *Figure 3*. We can begin to get some idea about the rate of reaction and its variation with time by considering these concentration profiles. First, it is apparent that the surface concentration gradient, $(\partial c/\partial x)_0$, decreases steadily with time, as more material close to the surface is consumed and the depletion of concentration moves further out into solution. From our initial discussion of diffusion we may recall that the transport rate is directly proportional to the concentration gradient, as formally described by Fick's first law, so the concentration profiles of *Figure 3* give an indication of the steady decay of surface reaction rate with time. A formal expression describing the variation of reaction rate with time is readily obtained by differentiation of equation 15 with respect to

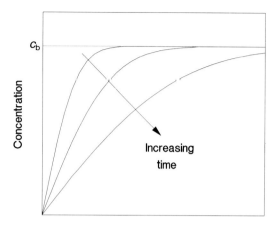

Figure 3. Concentration profiles in a stationary solution at a surface where reaction takes place linked to diffusional mass transport and their dependence upon time; curves are derived from the Cottrell equation, equation 16.

x with substitution of $x = 0$ to give the concentration gradient at the surface. This is then substituted into the flux expression of Fick's first law, equation 13, to give the Cottrell equation:

$$j_D = \left(\frac{D^{1/2}}{\pi^{1/2} t^{1/2}} \right) c_\infty \quad (16)$$

where j_D is the diffusion controlled flux per unit area. Thus, we have progressed from a visual concept of diffusion as a random process into a formal mathematical description, by means of which we have been able to generate the required visual picture of the state of the system in terms of concentrations and surface reaction rates. Further examples of the application of Fick's laws are given in Section 4.

3.4 Fundamentals of convective mass transport

Convection refers to the movement of the solution as a whole under the influence of an external mechanical force and not just to the random motion of the components within it, though the random diffusive processes will still occur and be relevant to the overall rates of mass transport within a convective system. Where reaction occurs at a surface, convection can be employed to enhance the rate of mass transport by replacing, with fresh solution, the solution close to the surface which has been depleted of reactant. It can also be employed to maintain the reaction at a constant rate provided that steady flow is maintained. This has important implications for biosensor applications in that it can provide first an enhancement of sensitivity and second a stable and controlled response rather than the decaying response we have seen for the diffusion-controlled example discussed earlier.

The mathematical description of mass transport by convective diffusion requires the addition of a further term to equation 14 to take account of the motion of the solution. This additional term simply describes the concentration change associated with the replacement, due to flow, of solution over a particular region of space within a stationary coordinate system, by fresh solution at a different concentration. Let us consider a small volume element of solution moving in the x direction at a velocity v_x. The distance travelled, δx, in a short time, δt, will be given simply by $\delta x = v_x \delta t$, the velocity multiplied by time. If there is a concentration gradient, $\delta c/\delta x$ in the solution then movement of the distance, δx, will lead to the replacement of solution at a point within the stationary coordinate system by solution at a concentration increased by the amount δc. The concentration change at that point in the time, δt, will therefore be given by $\delta c = v_x(\delta c/\delta x)\delta t$ and so, including this into equation 14 as an additional contribution to the rate of concentration change, we have:

$$\frac{\partial c}{\partial t} = D\left(\frac{\partial^2 c}{\partial x^2}\right) - v_x\left(\frac{\partial c}{\partial x}\right) \quad (17)$$

which describes the rate of change in concentration at a point in space due to convective diffusion.

In order to apply equation 17 to a real problem we require knowledge of the solution velocity, v_x, throughout the relevant region of space and the way the velocity varies with the position, x. This will depend upon the particular type of flow arrangement employed and a complete description of each of the different flow configurations likely to be encountered is beyond the scope of this chapter. However, it is worth while considering the principles behind the problem before stating some of the general features of flow systems commonly encountered.

The fundamental fluid property controlling the flow characteristics is viscosity, which effectively reflects the resistance of the solution towards sliding motion against itself. This resistance to motion of a fluid against itself is formally described by Newton's law of viscosity (10). For flow over a stationary surface, solution immediately adjacent to the surface will be stationary whilst the next layer of solution will move a little with respect to it, according to its viscosity. For a given force, each subsequent layer of solution moves a little with respect to the previous layer such that the velocity of the solution with respect to the stationary surface steadily increases with increasing distance from it. Thus, an essentially stagnant layer of solution exists close to the surface whilst fresh solution moves more freely further from it.

Where reaction takes place at the surface and species are destroyed or generated, they must diffuse through this stagnant layer, typically called the diffusion layer,

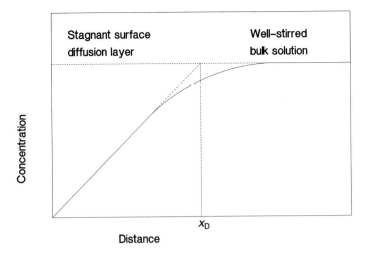

Figure 4. The concentration profile developed under conditions of convective mass transport in stirred solution.

between the surface reaction site and the well stirred bulk of solution. With the constant replenishment, from the flowing solution, of material consumed as a result of the surface reaction, a steady-state concentration profile across the diffusion layer, as illustrated in *Figure 4*, can then be achieved in which the rate of surface reaction is in balance with the rate of mass transport. This is in contrast with the transient behaviour observed for the stationary diffusion case described earlier. Mass transport by convective diffusion may therefore be characterized by a time-independent rate constant, k_D, with units of m sec^{-1}, with the reaction rate being described in terms of the transport flux, j_D, in moles per unit area, according to:

$$j_D = k_D(C_\infty - C_0) \tag{18}$$

Where C_∞ and C_0 refer to the bulk solution and surface concentrations, respectively.

It is often convenient to think in terms of the diffusion layer thickness, x_D, which is effectively the thickness of the stagnant surface layer across which material must diffuse between the well stirred bulk of solution and the surface. This diffusion layer thickness is related to the mass transport rate constant according to:

$$k_D = \frac{D}{x_D} \tag{19}$$

This effectively implies an abrupt break between the stagnant and well stirred regions of solution, with a correspondingly abrupt change in the concentration profile, as indicated by the dashed line in *Figure 4*. In reality the two regions merge and the concentration profile changes smoothly but the diffusion layer concept remains a useful one, particularly when considering coupled reactions close to the surface. The idea of the mass transport rate constant and diffusion layer thickness can also be usefully applied to the description of the transient flux in a stationary solution, with k_D given by the bracketed term in equation 16, though in this case these will vary with time, as can be seen from *Figure 3*.

The magnitude of the mass transport rate constant, k_D, will be dependent upon both the geometry of the flow arrangement employed and the rate of flow. It is derived for any particular convective system by solution of equation 17, having substituted the appropriate value for the convective velocity, v_x, and set the time differential, $\partial c/\partial t$, to zero, as appropriate for the steady state. The convective systems most commonly encountered in biosensor research and development are; first, the rotating disk system (12,13), in which the spinning action of the disk draws solution towards and across its surface, defining a mass transport rate constant which is dependent upon the rotation speed and is uniform across the disk surface; second, the rectangular or tubular channel system (14), for which the mass transport rate constant is both a function of flow rate through the channel and distance along the reactive surface within the channel, this latter aspect reflecting the increase in reactant depletion as the solution progresses downstream; third, the wall jet (15) and impinging jet (16) systems, in which solution issues from a nozzle, tangentially, against the surface of interest and spreads out radially, the mass transport rate constant again being dependent upon the solution flow rate. The detailed description of these systems with the exact functions describing their mass transport rate constants is beyond the scope of this chapter and can be found in the references above. For our current purposes it is sufficient to note the general features of convective systems and, in particular, that the convective diffusion transport rates can be described by the surface flux in terms of the mass transport rate constant, k_D, according to equation 18. When analysing complete reaction schemes, involving coupled mass transport and reaction, this rate constant can be employed in a manner analogous to that for any rate constant in a multistep homogeneous solution reaction.

3.5 Transport due to migration

Migration refers to the movement of charged species under the influence of an applied potential gradient. In an electrochemical system, it represents the mechanism by which charge flows through solution between the two electrodes at which ionic charge is created and destroyed by electron transfer, thereby maintaining the necessary charge balance. However, its contribution to the overall rate of mass transport of an electrochemically generated species is generally of no importance and need not concern us. This is due to the presence, in most cases, of a large excess of other ionic species, typically referred to as an inert supporting electrolyte, which is primarily responsible for the passage of the required charge and depresses any potential gradients which might otherwise arise in the solution. Thus, convection and diffusion are typically the only significant mechanisms of mass transport in the solutions currently of interest to us.

4. Coupled transport and reaction processes: analysis of complete systems

Now that the basic kinetic and equilibrium components of biosensors have been considered, a description of the complete system can be formulated. Typically, this will start with the mathematical description of each of the kinetic processes involved, perhaps in the form of a differential equation, along with the boundary conditions. From this, an algebraic expression might be derived, relating the quantity of interest, the analyte concentration, to the quantity measured. The latter will be perhaps an enzyme catalysed reaction product to the analyte, a current derived from it, or a complex formed between the analyte and the biosensor's receptor surface. This relationship might then be either represented graphically to illustrate the predicted performance characteristics of the system, compared with experimental data, or employed in the determination of the various kinetic parameters from experimental data. Where an algebraic solution for the relationship between analyte concentration and output signal is precluded by the complexity of the system a numerical solution of the problem might be necessary, from which a graphical representation of the system's performance can again be constructed.

Before commencing formulation of such a description, it is worthwhile considering its purpose. In an area of applied research such as biosensors, the primary roles of theory are first, to provide a framework within which the experimentally observed performance of a system can be understood and optimized and second, to assess which might be the best approach to the provision of new and improved systems. Whilst an exact solution of the appropriate equations to yield the complete system description will always be attractive, it should be remembered that the complexity of the problem will often preclude this. Simplifications may need to be made, perhaps by reference to limiting cases. In many instances, simplified treatments will be of more use in aiding an understanding of the system especially where more complete descriptions are so complex as to obscure the significance of important factors affecting system performance. This can be particularly true where numerical solution is the only route to a complete description. It will always be necessary experimentally to prove a system and a simplified theoretical treatment can often guide this process more effectively than a complex one. Understanding is often the key to effective practical research and it should be the aim of theory to provide this rather than absolute precision in the description of a system.

A number of examples have been chosen to illustrate how the provision of a theoretical description of biosensors might be approached and to illustrate a number of features that are commonly encountered. These features include, for example, the steady-state kinetic analysis involving mass and flux balance conditions, the use of approximate treatments and of limiting cases to provide simplification of the

Theoretical methods for analysing performance

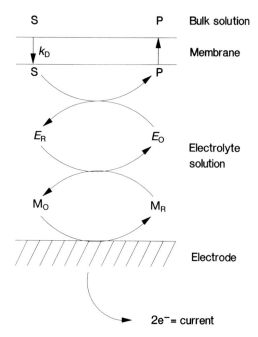

Figure 5. Schematic representation of the mediated amperometric biosensor, showing the kinetic processes involved.

problem, the concept of the rate determining step and the use of dimensionless variables in the formulation of a problem and characteristic dimensionless parameters in the final description.

4.1 Steady-state analysis of mediated amperometric systems

As a first example of the application of theory to the analysis of real systems we will consider the case of the mediated amperometric device, an example of considerable interest in biosensor research. We will start by defining the different elementary steps involved and their rates. Then, by straight forward algebraic manipulation, we will derive an expression describing the output signal in terms of the various kinetic parameters characterizing the system and the substrate concentration which it is the purpose of the device to measure. The system is illustrated schematically in *Figure 5* and is comprised of an electrode which monitors the mediator and provides the measurement signal, the electrolyte layer which contains the mediator and the enzyme and, finally, an outer membrane which serves to control diffusional mass transport to the electrolyte layer as well as to physically constrain the electrolyte close to the sensing electrode. The reaction sequence upon which the system is based can be summarized as follows:

$$S + E_O \underset{k_{-1}}{\overset{k_1}{\rightleftharpoons}} E_O S \tag{20a}$$

$$E_O S \overset{k_2}{\rightarrow} P + E_R \tag{20b}$$

$$E_R + M_O \overset{k_e}{\rightarrow} E_O + M_R \tag{20c}$$

$$M_R \rightarrow M_O + e^- \tag{20d}$$

This represents the simplest reaction sequence for the particular case of the mediated amperometric biosensor. More complete descriptions of the various approaches to amperometric biosensing are available elsewhere (8,17). Equations 20a and 20b represent the oxidation of substrate to product, according to Michaelis–Menten kinetics, by the oxidized enzyme, E_O, to give the reduced form of the enzyme, E_R. Equation 20c represents the re-oxidation of the reduced enzyme by oxidized mediator, M_O, giving rise to the reduced mediator, M_R. Equation 20d represents the re-oxidation of the mediator at an electrode which gives rise to the current upon which the measurement is based. The rate constants k_1, k_{-1} and k_2 are those for Michaelis–Menten kinetics, as described in Section 2.4, and k_e is the bimolecular rate constant for mediator reaction with enzyme. The reactions described by equation 20 are the chemical processes which take place in the enzyme layer and these are linked to the transport step by which the substrate is brought across the membrane from bulk solution to the electrolyte layer. We must therefore include this transport step, characterized by the rate constant $k_D = D/l_M$ where l_M is the membrane thickness, within the overall reaction sequence.

Having defined each of the elementary steps involved we can proceed to write down the rates (in this case the fluxes, j, per unit area) for each of them and the net rates of formation of the various intermediates. To do this will require the use of approximations in order to produce a straightforward and readily comprehensible description of the system. First we will employ the steady-state assumption, the use of which in the description of homogeneous solution enzyme kinetics has been described in Section 2.4. In this case, a steady state is established in which the rates of the different steps occurring in the electrolyte layer are balanced with one another and with the mass transport of reactant and product to and from it. Note that this steady-state assumption requires that the sensing reaction causes negligible change in the bulk solution concentrations of reactant and product, a condition which is readily fulfilled though should not simply be taken for granted in all cases. In addition, further simplifying assumptions regarding the reactions in the electrolyte layer will be employed. We will assume that the concentrations of the various species across the electrolyte layer behind the membrane are uniform. That is, no concentration polarization of the type discussed in Section 3 occurs. In this case, the fluxes for the reactions in the electrolyte layer can be written simply as the usual rate expression for the reaction multiplied by the electrolyte layer thickness, l. In fact, with substrate arriving at the outer surface of the electrolyte layer and being depleted within it, its concentration will tend to be non-uniform, being higher at the outer surface. However, concentration polarization will be minimal and the approximation valid provided the electrolyte layer is sufficiently thin so that diffusion times across it are short and that the diffusion rates are relatively rapid compared with the enzyme reaction rates. Further, we will assume that there is an excess of mediator and that the electrode regenerates M_O from M_R sufficiently rapidly that the concentration of M_O is effectively that of the total mediator $[M]$. This latter condition can be arranged in practice by ensuring that first, the re-oxidation occurs rapidly at the electrode, second, the electrolyte layer is thin so that the diffusional mass transport of mediator across it to the electrode is also rapid, and third, the mediator concentration is sufficiently in excess of the enzyme concentration.

It is to be stressed that the treatment will be applicable only where these conditions are satisfied and that, where these conditions are not satisfied, a more complex treatment (18) will be required, taking into account the concentration polarization in a manner analogous to that illustrated for the non-product consuming potentiometric biosensor in Section 4.2.2. The basis of this treatment is described in Section 5.1.1. We can assess the validity of our assumptions by considering the relative rates of transport and enzyme kinetics. The characteristic diffusion time across a thin layer, considered in Section 4.3, can typically be represented by $\tau = l^2/D$ where l is the layer thickness and D is the diffusion coefficient, the latter having a value of the order of 10^{-9} m^{2}sec^{-1} for a typical substrate. So, if $l \sim 10^{-5}$ m then we have $\tau \sim 10^{-1}$ sec and an effective rate constant for transport of substrate across the membrane of 10 sec^{-1}. As for the enzyme catalysed reaction, this cannot exceed the initial association rate of enzyme with substrate, with a pseudo-first-order rate constant, $k_1[E]$, with respect to the substrate. Taking glucose oxidase as an example, we have $k_1 \sim 10^4$ M^{-1}sec^{-1} and so, provided $[E] < 1$ mM we find $k_1[E] < 10$ sec^{-1} and our assumption that transport across the layer is rapid compared with reaction in it will be valid. The above considerations were applied to substrate and similar ones will apply to the mediator and enzyme.

Having satisfied ourselves of the validity of our approximate model we can proceed to write the flux equations for the various steps:

$j = k_D([S]_\infty - [S]_O)$ transport rate across the membrane   (21a)

$j = l(k_1[E]_O[S] - k_{-1}[E_OS])$ net formation rate of the complex   (21b)

$j = lk_2[E_OS]$ rate of the complex decomposition to give product and reduced enzyme   (21c)

$j = lk_e[E_R][M_O]$ rate of regeneration of oxidized enzyme and reduced mediator   (21d)

The simplification of the steady-state assumption allows us to set all these fluxes as equal and this flux must also equate with the current for regeneration of the oxidized mediator at the electrode, the measured signal, according to:

$$i = nFj \tag{22}$$

In addition, by assuming uniformity of concentration across the electrolyte layer we have been able to write down straightforward algebraic expressions for the flux as simply the uniform homogeneous solution reaction rate multiplied by the layer thickness, and have thereby avoided the need for solving differential equations.

Equations 21a–d represent a system of four simultaneous equations in five unknowns, the flux and the concentrations of the various intermediates, and we require from them a solution for the flux in terms of the bulk substrate concentration, $[S]_\infty$ and the various kinetic parameters. We can introduce a fifth equation defining the sum of the concentrations of the various enzyme intermediates:

$$[E_T] = [E_O] + [E_O S] + [E_R] \tag{23}$$

where $[E_T]$ is the known initial concentration of enzyme added, and we can now proceed to solve the system of five equations in five unknowns, eliminating the unknown concentrations of the intermediates to provide the final required relationship. The solution, given in a reciprocal form, as employed earlier in the description of Michaelis–Menten kinetics, is:

$$\frac{1}{j} = \frac{K_M}{lk_1[E_T]\left([S]_\infty - \frac{j}{k_D}\right)} + \frac{1}{lk_2[E_T]} + \frac{1}{lk_e M[E_T]} \tag{24}$$

The details of its derivation are left to the interested reader. We can now attempt to obtain from this expression some understanding of the behaviour of the system. If we wish to use the theoretical description in a purely predictive manner we might substitute values for the various kinetic parameters into a rearranged form of equation 24 and generate curves showing the flux as a function of substrate concentration. It we wish to interpret experimental data or gain an understanding of the behaviour of the system without a significant effort in computation and graph plotting, it may be more advantageous to retain the double reciprocal form but, because of its non-linearity in the flux, j, we will be forced to consider limiting cases which are linear. However, the double reciprocal form has already separated out the different terms to some degree. We can see that the first term on the right hand side in equation 24 is describing in some way the effect of the Michaelis–Menten type enzyme kinetics upon the reaction flux, though it has become linked with the mass transport process for substrate across the membrane. The second term on the right hand side describes the decomposition of the enzyme substrate complex into products and the third term describes the electron transfer reaction with mediator that regenerates the oxidized form of the enzyme. These then are the possible rate determining steps in the overall reaction sequence.

Theoretical methods for analysing performance

To gain a better understanding of the system from the description of equation 24 we can consider two limiting cases. First we have the case where the rate of decomposition of the enzyme−substrate complex into products and the mediated regeneration of the oxidized enzyme are very rapid compared with the preceding steps such that the reciprocals of these rates, the second and third terms on the right hand side of equation 24, are correspondingly small and their contribution to the control of the overall rate can be neglected. Dropping these terms and rearranging equation 24 gives:

$$\frac{1}{j} = \frac{K_M}{lk_2[E_T][S]_\infty} + \frac{1}{k_D[S]_\infty} \qquad (25)$$

in which the first term on the right hand side represents the reciprocal of the rate expression for enzyme kinetics and the second term represents the reciprocal of the rate of mass transport across the membrane. These then are the two possible rate determining steps considered in this limiting case. We can now consider what this tells us about the performance of a system consistent with this limiting case. First, it shows that the flux from which the measured signal is obtained is directly proportional to substrate concentration. This is an operational characteristic of value in a biosensor which one might specifically seek and be able to engineer through the kinetic understanding of the system. Second, it shows how, by varying either the enzyme concentration, $[E_T]$, or membrane thickness and hence k_D we can investigate the contributions of the rates of the two different steps to the overall rate. This information is not of purely academic interest as the variation of rate with enzyme concentration is indicative of the decrease in output that would occur with ageing if the enzyme denatures with time. To avoid this loss of response it would be necessary to ensure that rate control is dominated by transport across the membrane.

The other limiting case to be considered is where one or other of the enzyme catalysed reaction steps or regeneration of oxidized enzyme by the mediator is not rapid compared with the preceding mass transport step. Hence the second and third reciprocal terms on the right hand side of equation 24 are not negligible. In this case there will be an appreciable concentration, $[S]_O$, of substrate in the electrolyte layer and the transport flux, as given by equation 21a, will be significantly less than the maximum conceivable transport flux, were this concentration in the electrolyte layer zero, i.e. $j = k_D([S]_\infty - [S]_O) << k_D[S]_\infty$. Then, with $j/k_D << [S]_\infty$, we can drop the j/k_D term in the first reciprocal term on the right hand side of equation 24. Effectively, what we are saying is that the transport of substrate across the membrane is rapid compared with one or more of the subsequent reaction steps; mass transport is not the rate determining step and the supply of substrate to the electrolyte layer in no way limits the overall reaction rate. The substrate concentration within the electrolyte layer will then remain effectively at its bulk solution value and equation 24 reduces to:

$$\frac{1}{j} = \frac{K_M}{lk_2[E_T][S]_\infty} + \frac{1}{lk_2[E_T]} + \frac{1}{lk_e[E_T][M]} \qquad (26)$$

Each of the reciprocal terms on the right hand side of equation 26 again represents the different rate determining steps in the overall reaction. The first two terms describe the enzyme kinetics, exactly in the manner discussed in Section 2.4 for Michaelis−Menten kinetics. They are completely analogous to the terms in the Lineweaver−Burke plot, given by equation 11, differing only in the inclusion of the layer thickness, l, in the surface flux expression and they represent rate control by the enzyme−substrate complex formation and catalysed reaction step respectively. The third term represents the contribution to rate control by the reaction of reduced enzyme with mediator to regenerate the oxidized enzyme. Each term shows a different dependence upon the three concentration variables at our control, that of the substrate, enzyme and mediator. To probe the system experimentally we can keep two of these concentrations constant, change the third and investigate its effect on the observed flux. Then, by making double reciprocal plots and measuring slopes and intercepts from them, we can determine the characteristic rate for each step.

In summary then, we have seen how, by using the steady-state assumption and a further, carefully chosen and justified approximation, a description of the response of a mediated amperometric biosensor can be obtained by equating the fluxes of the various steps in the overall reaction and then solving the resulting set of simultaneous equations. The convenience of the double reciprocal representation of the reaction flux, due to the manner in which this separates out the different possible rate determining steps, has been illustrated and its application to the experimental determination of values for the various kinetic parameters characterizing the system has been described. The consideration of limiting cases as a means to overcome the non-linearity in the more complete description has also been illustrated. This allows an analysis of experimental data in a straightforward manner by means of simple linear plots. The interested reader is referred to further accounts (17,18) of the theoretical description of amperometric biosensors and methods of data analysis.

4.2 Immobilized enzyme layers with non-consuming reaction product detection

Enzyme layers, immobilized directly upon the sensing surface, have been employed extensively (8,19 and references therein) in potentiometric biosensors. In contrast to amperometric devices these do not consume product, which is lost from the sensor surface by mass transport alone. A considerable number of theoretical descriptions of this type of system have been given, with varying degrees of complexity and precision, and they provide an excellent opportunity to illustrate many pertinent features.

The system is illustrated schematically in *Figure 6* and consists of a sensor surface, lying at $x = 0$ in the appropriate coordinate system, coated with an immobilized enzyme layer of thickness l. Beyond this lies the transport boundary layer, perhaps controlled by a membrane or defined convective flow regime, with a characteristic mass transport rate constant, k_D. Beyond the transport boundary layer the concentration of analyte, the enzyme's substrate, S, has a defined value, $[S]_\infty$, and that of the product, P, is taken to be zero. Note that these bulk concentrations of

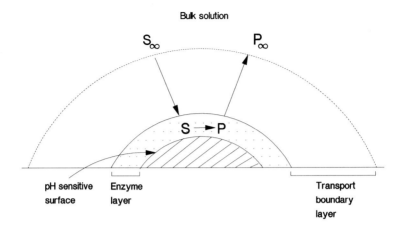

Figure 6. Schematic representation of the potentiometric biosensor employing an immobilized enzyme layer.

substrate and product represent two basic boundary conditions of the system. Enzyme-catalysed reaction occurs in the immobilized enzyme layer and, for the present example, will be considered to follow simple Michaelis−Menten kinetics, described by equation 11. This reaction depletes the substrate and generates the product in the enzyme layer, creating the concentration gradients which drive the mass transport process. The product is detected potentiometrically, without being consumed, at the sensor surface. As considered earlier for the amperometric system, a steady state is established, in which the rate of reaction in the immobilized enzyme layer is balanced by mass transport of reactant and product to and from it. This is the basic definition of the system upon which a number of theoretical treatments are based. Several of these are considered below in order of increasing complexity.

4.2.1 Simplified treatment: uniform concentration in the enzyme layer

As with the previous amperometric example, a simplified treatment can be obtained by assuming that the concentrations of species throughout the reaction layer, in which the enzyme catalysed reaction proceeds, are uniform and that the reaction within it is linked to a separate external mass transport process. In fact, the reaction layer is of finite thickness and the restricted rates of diffusional mass transport between the exposed outer surface and the interior of the immobilized enzyme layer, throughout which reaction takes place, will always lead to some degree of concentration polarization and hence a breakdown in uniformity across it. The description will therefore be approximate, its accuracy being dependent upon the thickness of the enzyme layer and the relative rates of mass transport and of the enzyme catalysed reaction. The usefulness of the model lies in its simplicity resulting from the elimination of the need to solve the differential equation of Fick's second law coupled, in this example, with Michaelis–Menten kinetics.

We are concerned with three processes, the transport flux, j_D, of substrate towards the surface which is described in terms of the mass transport rate constant, k_D, according to equation 18, the enzyme layer reaction flux, j_R, described by the Michaelis–Menten rate expression, equation 11, multiplied by the layer thickness, l, to give the flux per unit area and finally the transport flux of product away from the surface. We therefore start by writing down these basic flux equations from which the relationship between bulk solution concentration of substrate and surface concentration of product are derived. These are as follows:

$$j_D = k_D([S]_\infty - [S]_l) \tag{27a}$$

$$j_R = \frac{lk_2[E][S]_l}{K_M + [S]_l} \tag{27b}$$

$$j_D = k_D[P]_l \tag{27c}$$

where $[S]_l$ and $[P]_l$ are the substrate and product concentrations throughout the reaction layer. In the steady state these fluxes will be equal making equations 27a–c a set of three simultaneous equations in the three unknowns, j, $[S]_l$ and $[P]_l$, the last of which is of interest to us. We can therefore proceed much as in the preceding section, eliminating unknowns by equating flux equations until we arrive at an expression for the required unknown. So, equating equations 27a and 27b gives an expression for $[S]_l$:

$$\frac{lk_2[E][S]_l}{K_M + [S]_l} = k_D([S]_\infty - ([S]_l) \tag{28}$$

which has eliminated the unknown flux. Rearrangement of equation 28 will give an expression for $[S]_l$ in terms of the various rate parameters and the defined bulk concentration, $[S]_\infty$, in a quadratic form which can be solved by use of the usual quadratic formula. However, it may be more instructive to look initially at the two limiting cases for which the algebraic expressions for $[S]_l$ are simpler and their meaning more readily understood.

For the first limiting case, where $[S]_l << K_M$, dropping the $[S]_l$ term in the denominator of the Michaelis–Menten rate expression in equation 28 gives, after minor rearrangement:

$$[S]_l = \frac{1}{1 + \frac{lk_2[E]}{k_D K_M}} [S]_\infty \tag{29}$$

This may then be substituted for $[S]_l$ in the transport flux expression for substrate, equation 27a, which is then equated with the transport flux expression for product, equation 27c, to give the required relationship between the bulk substrate concentration and surface product concentration:

$$[P]_l = \frac{\frac{lk_2[E]}{k_D K_M}}{1 + \frac{lk_2[E]}{k_D K_M}} [S]_\infty \qquad (30)$$

The second limiting case to be considered is that where $[S]_l \gg K_M$. In this case, dropping the K_M in the denominator of the Michaelis–Menten rate expression in equation 28 gives an expression for the surface reaction flux which is independent of substrate concentration and equating this with the product transport flux gives:

$$[P]_l = \frac{lk_2[E]}{k_D} \qquad (31)$$

This expression describes the enzyme limited response at high substrate concentration and, when taken together with the previous limiting case, gives a simple description of the system's output response as a function of the various rate parameters of the system and the analyte concentration. By solving the quadratic derived from equation 28 the region in between those covered by the limiting cases can be defined. The behaviour of the system is illustrated in *Figure 7* which demonstrates how both sensitivity and operating concentration range vary with the values of the relevant rate parameter.

The above derivation has illustrated the use of the reaction layer model, and has again illustrated the relative simplicity of the steady state which provides a set of readily soluble simultaneous equations. The form of the resulting expression, equation 30, also serves to illustrate another important feature, the dimensionless parameter, in this case $(lk_2[E])/(k_D K_M)$, which frequently arises in theoretical descriptions. In this instance the dimensionless parameter reflects the ratio of surface reaction rate constant, $(lk_2[E])/K_M$, to transport rate constant k_D. Both rate constants have the same dimensions, that of a surface rate constant, m sec^{-1}, and the ratio is therefore without dimensions. All the important parameters which determine the system response are contained within this dimensionless parameter, these being the enzyme layer thickness, the enzyme concentration within it, the intrinsic catalytic efficiency of the enzyme and the mass transport rate constant. We can identify two types of limiting behaviour for the system, according to the value of the dimensionless parameter. For the first, where $(lk_2[E])/(k_D K_M) \gg 1$ and the rate constant for transport is slower than that for the surface reaction, the process is transport limited.

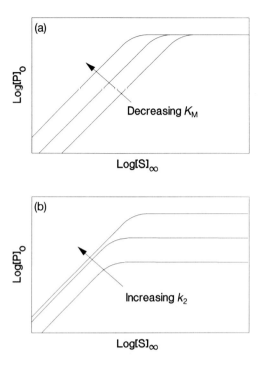

Figure 7. Response curves for the potentiometric biosensor employing an immobilized enzyme layer showing the different effects as regards improvements in sensitivity of (a) decreasing K_M and (b) increasing k_2.

Under these conditions substrate is converted to product as fast as it arrives at the surface and the optimum sensitivity is obtained, where $[P]_l = [S]_\infty$. For the second, where $(lk_2[E])/(k_D K_M) \ll 1$, the surface reaction rate is limiting. In this case the characteristic rate for conversion of substrate to product is slow compared with the transport of product away from the surface. The sensitivity is therefore reduced, in proportion to the ratio of rate constants. From the point of view of optimizing the performance of the system this dimensionless parameter can be employed to choose the various rate controlling parameters, for example the layer thickness, l, and enzyme concentration, $[E]$, in order to obtain maximum sensitivity. Consideration of dimensionless parameters in this way, to identify the various limiting behaviours of a system, can be extremely instructive and will be illustrated several times more in subsequent examples.

4.2.2 Concentration polarization and mass transport effects in the enzyme layer

The preceding reaction layer treatment was relatively simple to perform, requiring nothing more complex than algebraic rearrangement and the solution of simultaneous equations. To achieve this, transport across the enzyme layer was neglected and the resulting solution is therefore only approximate. If the effects of diffusional mass transport across the enzyme layer are to be included, Fick's second law of diffusion, given by equation 14, must be employed such that the theoretical analysis will involve solution of a differential equation. The problem is, however, quite straightforward and will serve to illustrate the solution of such a differential equation subject to boundary conditions.

We can start by solving the substrate concentration profile across the enzyme layer, in terms of the substrate concentration $[S]_l$ at the outer surface of the immobilized enzyme layer at $x = l$. Two processes effect change in the substrate concentration throughout the enzyme layer, the enzyme catalysed reaction and diffusional mass transport. Combining the relevant expressions for the two processes, according to equations 11 and 14, and setting the time differential to zero since we are concerned with the steady state, gives a differential expression describing the substrate concentration across the immobilized enzyme layer:

$$D_s \left(\frac{d^2[S]}{dx^2} \right) - \frac{k_2[E][S]}{K_M + [S]} = 0 \qquad (32a)$$

where D_s is the diffusion coefficient of the substrate. A similar expression describes the product:

$$D_p \left(\frac{d^2[P]}{dx^2} \right) + \frac{k_2[E][P]}{K_M + [P]} = 0 \qquad (32b)$$

where D_p is the diffusion coefficient of the product and the difference in sign between the expressions for substrate and product takes account of the consumption of the one and generation of the other by enzyme catalysed reaction. Complete solution of these equations is possible only by recourse to numerical methods and we are immediately forced to consider the two limiting cases, where either $[S]_l \ll K_M$ or $[S]_l \gg K_M$, to discover analytical solutions. These correspond to the cases of low and high substrate concentration where the reaction is limited by either substrate or enzyme concentration, as described in Section 2.4 in the discussion of Michaelis–Menten kinetics and encountered again in Section 4.2.1.

Taking first the case where $[S]_l \ll K_M$ equation 32a simplifies to give:

$$\frac{d^2[S]}{dx^2} = \alpha [S] \qquad (33)$$

where α is the enzyme loading factor defined by:

$$\alpha = \frac{k_2[E]}{K_M D_s} \qquad (34)$$

To solve this case we must refer to the boundary conditions, that is the state of the system and nature of the substrate concentration function at the boundaries of the region over which equation 33 applies. One is already defined, in that the concentration at the outer surface of the immobilized enzyme layer is a defined constant, i.e. at $x = l$, $[S] = [S]_l$. For the other, at $x = 0$, we note that this boundary is inert, the substrate is neither consumed nor generated there such that, with no reaction at the surface, there can be no reaction flux. Neither can there be any transport flux; the surface concentration gradient must be zero and we may write the two boundary conditions:

$$x = 0, \quad \frac{d[S]}{dx} = 0 \qquad (35a)$$

$$x = l, \quad [S] = [S]_l \qquad (35b)$$

The same boundary conditions will apply to the product, P, and it is often the case that $[P]_l \sim 0$. The solution of this problem, which is described in Section 5.1.1, gives the substrate and product concentrations across the enzyme layer:

$$[S] = \frac{\cosh(x\sqrt{\alpha})}{\cosh(l\sqrt{\alpha})}[S]_l \qquad (36a)$$

$$[P] = \frac{D_s}{D_p}[S]_l\left(1 - \frac{\cosh(x\sqrt{\alpha})}{\cosh(l\sqrt{\alpha})}\right) \qquad (36b)$$

Setting $x = 0$ in equation 36b gives the surface concentration of product in terms of the substrate concentration in the solution:

$$[P]_0 = \frac{D_s}{D_p}[S]_l\left(1 - \frac{1}{\cosh(l\sqrt{\alpha})}\right) \qquad (37)$$

This expression shows that, for the limit where $K_M >> [S]_l$, the surface concentration is directly proportional to the substrate concentration at the enzyme layer outer surface and from it response curves for the system, as a function of the relevant kinetic parameters of the system, can be constructed, As for the previous description, obtained by using the reaction layer treatment, this description of the system involves a dimensionless parameter, in this case, $l\sqrt{\alpha}$. This dimensionless parameter incorporates the important kinetic variables of the system and reflects the relative rates of the enzyme catalysed reaction in the immobilized enzyme layer and of diffusional mass transport across it. Curves, illustrating how the concentration profiles across the enzyme layer vary with the magnitude of the dimensionless parameter, can be constructed, as shown in *Figure 8*. These demonstrate how

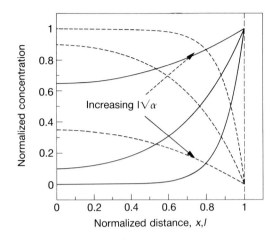

Figure 8. Concentration profiles of substrate (solid line) and product (dashed line) across an immobilized enzyme layer: curves generated using equation 36a/b.

increasing the rate of the enzyme catalysed reaction relative to the rate of mass transport leads to increased depletion of the substrate concentration, an increased product concentration at the surface and hence an increased sensitivity for the system.

Finally, we should consider the zero-order kinetic limit, where $[S] \gg K_M$. Under these conditions equations 32a and 32b reduce to:

$$D_s \frac{d^2[S]}{dx^2} - k_2[E] = 0 \tag{38a}$$

$$D_p \frac{d^2[P]}{dx^2} + k_2[E] = 0 \tag{38b}$$

These differential expressions are readily integrated, subject to the boundary conditions previously defined, to give expressions for the substrate and product concentrations:

$$[S] = [S]_l - \frac{k_2[E]}{SD_s}(l^2 - x^2) \tag{39a}$$

$$[P] = \frac{k_2[E]}{2D_p}(l^2 - x^2) \tag{39b}$$

From this it can be seen that the product concentration is dependent upon the relative rates of the enzyme limited reaction and diffusional mass transport, and is independent of substrate concentration. This behaviour is consistent with that derived earlier from the reaction layer treatment.

4.2.3 Concentration polarization in the enzyme layer linked with bulk solution mass transport

The reaction layer treatment (Section 4.2.1) considered mass transport from bulk solution linked with an approximate treatment of the reaction in the enzyme layer whilst the treatment in Section 4.2.2 treated the reaction in the enzyme layer more accurately but ignored the bulk solution mass transport process, assuming transport to be rapid compared with reaction. Where mass transport is not sufficiently rapid to ensure that the concentration at the interface between solution and the enzyme layer, $[S]_l$, is the same as the bulk, $[S]_\infty$, its effect can be brought into the latter description by reference to the requirements for flux balance and continuity of concentration (more precisely of activity given that different phases are involved, for which different activity coefficients might apply) at the boundary. The description obtained will be more precise than those given earlier, as well as a little more complex and is derived next.

Where concentration polarization in bulk solution occurs, the transport flux of substrate will be as given in the reaction layer treatment by equation 27a. In the steady state this transport flux from solution must equal the diffusional flux into the enzyme layer at $x = l$, on the inner surface of the enzyme layer. This latter flux in the enzyme layer, j_E, is proportional to the concentration gradient at the surface which is obtained by differentiation of the expression for the substrate concentration, equation 36a:

$$j_E = D_s \left(\frac{d[S]}{dx} \right)_{x=l} = D_s[S]_l \sqrt{\alpha} \tanh(l\sqrt{\alpha}) \tag{40}$$

Equating this with the bulk solution transport flux we can eliminate the unknown flux to give an expression for the concentration at the enzyme layer surface in terms of that in the bulk:

$$[S]_l = \frac{[S]_\infty}{1 + \dfrac{D_s \sqrt{\alpha} \tanh(l\sqrt{\alpha})}{k_D}} \tag{41}$$

This expression for $[S]_l$ may be substituted into the various expressions derived in the previous section to give the concentrations of substrate and product across the enzyme layer in terms of the bulk solution concentration. Most importantly, an expression for the product concentration at the sensor surface is obtained:

$$[P]_0 = \frac{D_s}{D_P}[S]_\infty \left(1 - \frac{1}{\cosh(l\sqrt{\alpha})}\right) \left(\frac{1}{1 + \dfrac{D_s\sqrt{\alpha}\tanh(l\sqrt{\alpha})}{k_D}}\right) \tag{42}$$

Theoretical methods for analysing performance

In this expression there are two important rate terms in parentheses which determine the behaviour of the system. The first includes the dimensionless parameter, $l\sqrt{\alpha}$, and describes the way the behaviour of the system varies with the relative rates of the enzyme catalysed reaction in the layer and the diffusional mass transport across it. The second term, which includes a different dimensionless term, $D_s\sqrt{\alpha}\tanh(l\sqrt{\alpha})/k_D$, describes the way the behaviour of the system varies with relative rates of the surface layer reaction and mass transport to it. Consideration of the physical meaning of these dimensionless terms and their influence upon the system's behaviour can be very instructive. Effectively, equation 42 demonstrates the same general principles as the earlier two simpler treatments, that is, variation of sensitivity with relative rates of mass transport and the enzyme catalysed reaction and dependence on the various rate parameters, but brings them together to give a more precise overall description.

4.2.4 Numerical solution of the general problem

Because of the non-linearity of the Michaelis–Menten kinetic term it has not been possible to find exact analytical solutions to the basic differential equations describing the system and reference to the limiting zero-order and first-order kinetic cases has been required. To obtain a complete solution, including the intermediate region between the two limiting cases, where $[S] \sim K_M$, numerical methods are required. A brief account of these is given in Section 5.2. The numerical solution will have features in common with the analytical solution. That is, it will involve the various dimensionless parameters of the system, for example, the kinetic parameter $l\sqrt{\alpha}$, and the dimensionless concentration, $[S]/K_M$. Typically, values for these parameters are chosen and the numerical solution then generates plots illustrating the system's behaviour for the range of values of these parameters employed. The general message conveyed by these plots as regards the system's behaviour will be the same as that given by the earlier, analytical, treatments but the precision with which the intermediate region, where $[S] \sim K_M$, is described will be much improved. Numerical solution will also allow for the introduction of other non-linear factors, for example product inhibition, the pH dependence of the enzyme kinetics and buffering effects where an acidic product is involved.

4.2.5 Summary of treatments of the immobilized enzyme layer based system

It has been the aim of the previous sections to illustrate the variety of approaches by which a theoretical description of the response of a non-product consuming biosensor can be obtained. Each has a number of features in common which are worth highlighting. In each case reference to mass and flux balance has been made in the formulation of the description and dimensionless variables have either arisen naturally in the solution or have been introduced during the formulation of the problem. The dimensionless variables reflect the relative magnitudes of the characteristic rates of the different possible rate determining steps. A number of possible rate determining steps have been identified. In the first instance the system may be limited by either mass transport or enzyme kinetics and, if limited by the latter, this can be either substrate concentration or enzyme concentration limited as is typically the case for Michaelis–Menten kinetics. The general message conveyed by each description is essentially the same. Where the approaches differ is in their complexity and degree of precision and it is up to the individual researcher to decide which is most appropriate to any given problem.

4.3 Response time of a diffusion membrane: solution of a time-dependent problem

So far we have considered the description only of the steady-state response of a system and the emphasis has been on those factors which control sensitivity. The time taken to reach the steady state, at which the measurement is made, can be of considerable importance, the response time being, in many cases, as important a criterion as sensitivity by which to judge a system's performance. As discussed in the preceding sections, diffusion membranes provide a useful means for controlling the system's steady-state response, ensuring in some cases linearity and in others optimum sensitivity. The determination of the response time associated with these diffusion barriers is well worth considering, being central to the effective operation of a number of biosensor devices. This will provide a useful example of the solution of a partial differential equation which describes a time-dependent problem. We will again see the appearance of a dimensionless parameter in the solution, in this case Dt/l^2, which characterizes the relaxation time of diffusion layers. A similar, thin layer diffusion problem relating to immunological binding at a surface provides a useful further example of a time-dependent problem.

4.3.1 Transmembrane diffusion

When a biosensor, such as the amperometric or potentiometric types described in the previous sections, is first introduced into the analyte solution there will be no analyte in the layer in which the enzyme reaction takes place and consequently no product for the device to measure. As the substrate diffuses into the enzyme layer the response will develop towards its steady-state value. The exact description of this time-dependent response of either the amperometric or potentiometric system is somewhat complex. However, in these cases, the membrane thickness will typically be chosen such that diffusion across it is the rate limiting process as this will provide both the optimum sensitivity and the most stable response. To a first approximation then, the relaxation time of the diffusion membrane will limit the response time of the overall system and we can obtain an estimate of the latter as a whole by consideration of the transmembrane diffusion process alone. The diffusional transport process across the membrane will be described by Fick's second law, equation 14, and to solve it we will need to define the relevant initial and boundary conditions. Taking, for example, the mediated amperometric case (Section 4.1), given that the reactions occurring within the solution layer contained by the membrane are rapid, the analyte will be consumed as fast as it arrives across the membrane such that the analyte concentration on the interior of the membrane will be effectively zero. Transport from bulk solution to the outer membrane surface will similarly be rapid compared with transport across it such that no concentration polarization in solution will occur and the bulk solution concentration of analyte will be maintained at the membrane. Thus the boundary conditions are that first, at the inner surface of the membrane, $x = 0$, the concentration at all times is zero and second, at the outer surface of the membrane, $x = l$, the concentration is at all times the bulk solution value. These boundary conditions may be formally written as:

$$t \geq 0, \quad x = 0 \qquad C(0,t) = 0 \qquad (43a)$$

$$t \geq 0, \quad x = l \qquad C(l,t) = C_\infty \qquad (43b)$$

where the concentration, C, is now a function of both distance and time, values of which are specified in parentheses. Next we need to define the initial condition, the state of the system at time $t = 0$. For this we have simply that the concentration is zero throughout the membrane prior to introduction of the analyte solution to its outer surface, which is formally written as:

$$t = 0, \quad 0 < x < l, \qquad C(x,0) = 0 \qquad (44)$$

The problem is now defined and can be solved using the Fourier method as described in Section 5.1.2. This gives an expression for the concentration at all points across the layer as a function of time:

$$C(x,t) = C_l \left[\frac{x}{l} + \sum_{n=1}^{\infty} \frac{2(-1)^2}{n\pi} \sin\left(n\pi \frac{x}{l}\right) \exp\left(-(n\pi)^2 \frac{Dt}{l^2}\right) \right] \qquad (45)$$

We are interested in the flux at the interior surface of the membrane, $x = 0$, and, more particularly, the surface flux at a given time compared with its steady state value, $j_{(t=\infty)} = DC_l/l$. Differentiating this expression for concentration with respect to distance and substituting $x = 0$ gives the surface flux which is then normalized with respect to the steady-state response to give the fractional response according to:

$$\frac{j_t}{j_\infty} = 1 + \sum_{n=1}^{\infty} 2(-1)^n \exp\{-(n\pi)^2(Dt/l^2)\} \qquad (46)$$

So, the state of the system, and the extent of its deviation from the ultimate steady-state response will be reflected by the size of the exponential summation term compared with 1. In this exponential summation term, which is a convergent series, the time variable appears as part of the dimensionless parameter, Dt/l^2, which describes how the relaxation time is a function of both the membrane thickness, l, and the diffusion coefficient, D. In general it will be found that the system will be

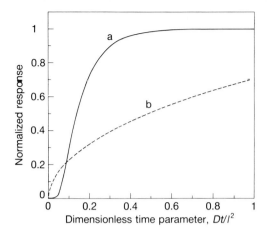

Figure 9. Time-dependent response curves for (a) the transmembrane diffusional flux, calculated from equation 46 and (b) binding at a receptor surface, calculated from equation 50.

significantly far from its steady-state position if $Dt/l^2 \ll 1$, and effectively at steady state when $Dt/l^2 \gg 1$. It is usual to consider the relaxation time, τ, to be given approximately by l^2/D. The precise form of the response can be found by substitution of values for Dt/l^2 into equation 46. Evaluation of the infinite sum presents no problem in this case as the first few terms or indeed the first term alone are dominant for values of the dimensionless time parameter, $Dt/l^2 > 0.1$, covering the time domain of primary interest. It is perhaps worth noting the form of the time-dependent response curve. This is illustrated in *Figure 9a* and shows the initial delay in the transport flux across the membrane as the transported species diffuses into it and the subsequent rise to the ultimate steady-state value for the transmembrane flux at longer times.

Thus, we have developed a simple treatment for determining the response time of a biosensor, limited by a diffusion membrane. The importance of the dimensionless parameter, Dt/l^2, in time-dependent diffusion problems has been illustrated and the use of the Fourier method in the solution of partial differential equations has been demonstrated.

4.3.2 Transport limited binding at a receptor surface

As a second example of the solution of a time-dependent problem, we can consider the binding of species at a receptor surface. In this case, rather than being interested in the concentration in solution at the sensor surface or the flux of material to it, as such, we are interested in the amount of material accumulated on the surface over a period of time. To find this we will need first to solve the relevant equation for concentration, which will serve to illustrate the use of the Laplace transform method in solution of partial differential equations. Then the surface flux can be found by

Theoretical methods for analysing performance

differentiation of the expression for concentration with respect to distance and the accumulated coverage after a given time determined by integration of the surface flux expression with respect to time.

We will consider the capillary fill type cell (20) comprising a thin solution film of thickness, l, confined between two plates, one of which is inert and the other being the receptor surface at which the binding reaction takes place. For simplicity we will confine ourselves to the case where the amount of binding species available in the solution is insufficient to cover the receptor surface completely, where the dissociation constant, K_d, is small compared with the concentration of the binding species and where the rate of binding at the surface is rapid. In this case (5), the surface coverage will be limited not by the surface binding reaction itself but by the amount of the binding species in the solution and the mass transport-controlled rate at which it reaches the receptor surface. As with the diffusion membrane example given in the previous section, we will consider diffusional mass transport according to Fick's second law, described by equation 14, and will begin by defining the problem in terms of the initial and boundary conditions for the system. The initial condition is quite straightforward, the surface coverage being zero and the concentration in solution being constant at its initial bulk solution value such that we have:

$$t = 0, \quad 0 < x < \infty, \qquad C(x,0) = C_i \qquad (47)$$

For the boundary conditions we have first that the concentration of the binding species at the receptor surface at $x = l$ is zero, given that the surface reaction is rapid compared with mass transport, and that its concentration gradient at the inert surface at $x = 0$ is zero, since it is neither generated nor consumed there:

$$x = l, \quad t > 0 \qquad C(0,t) = 0 \qquad (48a)$$

$$x = 0, \quad t > 0 \qquad \left(\frac{dC}{dx}\right) = 0 \qquad (48b)$$

The usual expression for diffusional mass transport, according to Fick's second law (equation 14), can be solved by the Laplace transform method, subject to the above initial and boundary conditions, as illustrated in Section 5.1.2, to give an expression for concentration across the solution layer and as a function of time which, normalized with respect to the initial concentration, has the form:

$$\frac{C_{(x,l)}}{C_i} = -\frac{4}{\pi} \sum_{n=1}^{\infty} \frac{(-1)^n}{(2n-1)} \exp\left[-\left(\frac{(2n-1)\pi}{2}\right)^2 \frac{Dt}{l^2}\right] \cos\left(\frac{(2n-1)n\pi x}{2l}\right) \qquad (49)$$

The time dependence is described by the exponential term containing the dimensionless time parameter, Dt/l^2, and the dependence upon distance across the chamber is described by the cosine term containing the normalized or dimensionless distance parameter, x/l. To find the surface coverage at a given time we must determine the rate of coverage as a function of time and then sum this over the time for which the accumulation of bound species has been proceeding. We have assumed a mass transport limit for the surface rate and the rate of coverage will therefore be the diffusional flux at the surface, according to Fick's first law as given by equation 13. If we differentiate equation 49 with respect to distance, x, to give an expression for the concentration gradient and then substitute $x = 0$ into the resulting expression we obtain the surface flux which represents the rate of binding as a function of time. Next we can sum this surface rate over the time of exposure to the binding species by integration with respect to time to find the surface coverage normalized with respect to the limiting surface coverage, $\gamma_{\lim}$, this limiting coverage being simply the total amount of binding species available in solution per unit area of binding surface, lC_i. The expression for normalized surface coverage obtained is:

$$\frac{\gamma(t)}{\gamma_{\lim}} = 1 - \sum_{n=1}^{\infty} \frac{8}{\pi^2(2n-1)^2} \exp\left[-\left(\frac{(2n-1)\pi}{2}\right)^2 \left(\frac{Dt}{l^2}\right)\right] \tag{50}$$

The form of the time-dependent response is illustrated in *Figure 9b* which shows the rapid initial rise and slower approach to the ultimate steady-state value at longer times. We can now use this expression to assess the likely performance of the capillary fill type system, in particular its sensitivity and response time in relation to the cell thickness, l. As with the previous example of a time-dependent problem, the time dependence is represented by the exponential summation term containing the dimensionless time parameter, Dt/l^2. From it we can determine the maximum thickness, l, of the capillary cell for which a chosen fraction response ($\gamma(t)/\gamma_\infty$) can be reached within a specified time, given the diffusion coefficient, D, of the binding species. It can be seen that, from the point of view of minimizing response time, the thinner the cell the better but the implications regarding sensitivity which decreases linearly with thickness, according to $\gamma_{\lim} = lC_i$, must be offset against this. Our model allows us to assess this quantitatively and predict the limits to the performance of the system.

In summary then, we have seen how we can solve the diffusion problems arising in the description of the time-dependent phenomena associated with biosensors by application of the Laplace transform or Fourier method, subject to the initial and boundary conditions relevant to the particular system under consideration. This description has involved a dimensionless parameter, in this case the dimensionless time variable, Dt/l^2, which compares the reaction time, t, with the characteristic relaxation time of the layer, $\tau \sim l^2/D$.

5. Mathematical methods

The theoretical description of biosensors derived earlier in this chapter has relied upon algebraic manipulation for the solution of simultaneous flux equations and the solution of differential equations which are required to describe the surface reaction coupled with mass transport to that surface. It has been assumed that the typical reader will at least be familiar with straightforward algebra. The main aim of this section is to illustrate the solution of differential equations with which many researchers may be somewhat less familiar. It will, however, be assumed that the reader is, to a certain extent, familiar with the terminology and meaning of differential equations if not so much with the practical matters of their solution. The breadth of the subject and the space available in this chapter do not permit a thorough and detailed account; mathematical texts (21,22) are available that do this and general accounts of the subject are given in standard texts (9,11) dealing with transport phenomena. Instead, an attempt has been made to give a reasonable account of what is involved, practically, by working through the solutions of some specific examples of relevance to biosensors. It is hoped that this will both aid the reader's understanding of these specific cases and assist, by way of example, those tackling the solution of similar problems.

We can identify two approaches to the solution of differential equations. The first is analytical, in which an algebraic expression is obtained, relating one variable, for example concentration, to the others upon which it depends, for example time and distance. The second is numerical, in which a value for the one variable relating to specific chosen values of the variables upon which it depends is calculated by numerical means, a separate calculation or set of calculations being required for each new value of the specified variable. Generally, analytical solutions are to be preferred where it is possible to obtain them, since they provide an exact description of the relationship between the relevant variables, applicable over a wide range of values. This relationship can indicate the behaviour of a system by virtue of the mathematical functions it contains. However, the complexity of many differential equations prevent their exact analytical solution such that it is often necessary to resort to numerical methods.

5.1 Analytical solution

5.1.1 Solution of an ordinary differential equation

The case of coupled diffusional mass transport and catalysed reaction in an immobilized enzyme layer has been considered in Section 4.2 and provides a suitable example for the illustration of the solution of an ordinary differential equation. The process of solution consists of first, mathematically defining the problem, that is, writing the appropriate differential equation for the system under consideration; second, provision of a general solution to this differential equation by inspection and substitution of an appropriate function and, finally, evaluation of the constants in the general solution by reference to the boundary conditions.

As discussed in Section 4.2, the basic differential equation describing the system at the lower concentration limit is:

$$\frac{d^2[S]}{dx^2} = \alpha[S] \tag{51a}$$

with a corresponding equation for product:

$$\frac{d^2[P]}{dx^2} = -\alpha[S] \tag{51b}$$

where α is the enzyme loading factor defined by:

$$\alpha = \frac{k_2[E]}{K_M D_s} \tag{52}$$

For the non-consuming, potentiometric sensing system, equation 51 is subject to the boundary conditions as follows:

$$x = 0 \qquad \frac{d[S]}{dx} = 0 \tag{53a}$$

$$x = l \qquad [S] = [S]_l \tag{53b}$$

From the knowledge of simple differentiation we know that $d\{\exp(x)\}/dx = \exp(x)$ and solutions of differential equations of the form of equation 51 will typically involve exponential functions. Consideration of the exponential functions fitting this equation provides the general solution to it:

$$[S] = A\exp(x\sqrt{\alpha}) + B\exp(-x\sqrt{\alpha}) \tag{54}$$

where A and B are constants. To determine these constants we refer to the boundary conditions for the system. To apply the first boundary condition of equation 53a we first differentiate equation 54 to give an expression for $d[S]/dx$:

$$\frac{d[S]}{dx} = \sqrt{\alpha}\{A\exp(x\sqrt{\alpha}) - B\exp(-x\sqrt{\alpha})\} \tag{55}$$

Then, substituting $x = 0$ and $d[S]/dx = 0$ in this differential gives $A = B$. The expression for [S] then becomes:

$$[S] = 2A\cosh(x\sqrt{\alpha}) \tag{56}$$

The other boundary condition is applied by setting $x = l$ and $[S] = [S]_l$ to give a value for the constant:

$$2A = \frac{[S]_l}{\cosh(l\sqrt{\alpha})} \tag{57}$$

Theoretical methods for analysing performance

and substituting this into equation 56 gives an expression for the substrate concentration across the enzyme layer:

$$[S] = \left(\frac{\cosh(x\sqrt{\alpha})}{\cosh(l\sqrt{\alpha})} [S]_l \right) \qquad (58)$$

In fact it is the product concentration, specifically that at the sensor surface at $x = 0$, that is of interest. To find this from the above expression for the substrate concentration we can make use of a further property of the steady-state system, that of mass balance across the enzyme layer. If we simply add the two basic differential expressions describing substrate and product, equations 51a and 51b, the kinetic term due to the enzyme catalysed reaction disappears giving:

$$D_s \left(\frac{d^2[S]}{dx^2} \right) + D_p \left(\frac{d^2[P]}{dx^2} \right) = 0 \qquad (59)$$

Now if we integrate this we find that $D_s(dS/dx) + D_p(dP/dx)$ is some constant and this applies throughout the enzyme layer. The differentials, $D_s(dS/dx)$ and $D_p(dP/dx)$ at $x = l$ represent the diffusional fluxes of substrate in and product out of the enzyme layer and, since material is neither created nor destroyed in the enzyme layer, only converted from S to P, their sum must be zero. Therefore if $D_s(dS/dx) + D_p(dP/dx)$ is zero at $x = l$ and takes the same constant value throughout the layer it must be zero everywhere. Note we would have found the same had we considered the boundary at $x = 0$ for which the boundary condition, defined earlier, is that the fluxes of both S and P are zero. Now if we integrate again we find the mass balance condition for the layer:

$$D_s[S] + D_p[P] = \text{constant} = D_s[S]_l + D_p[P]_l \qquad (60)$$

This mass balance condition is an important property of the steady state and is often useful in relating the concentrations of different species. So, with $[S]_l$ and $[P]_l$ defined and $[S]$ given by equation 58, substituting into equation 60 we obtain an expression for the product concentration:

$$[P] = \frac{D_s}{D_p}[S]_l \left(1 - \frac{\cosh(x\sqrt{\alpha})}{\cosh(l\sqrt{\alpha})} \right) \qquad (61)$$

which is effectively the complement of the substrate concentration. Setting $x = 0$ gives the surface concentration of product in terms of the substrate concentration in the solution:

$$[P]_0 = \frac{D_s}{D_p}[S]_l \left(1 - \frac{1}{\cosh(l\sqrt{\alpha})} \right) \qquad (62)$$

The application of this expression to the understanding of the performance of the potentiometric biosensor has been discussed in Section 4.2.2.

The form of the above solution of the basic differential equations describing the system, equations 51a and 51b, is dependent, not only on the basic equation but also on the boundary conditions which have been employed to determine the constants appearing in the general solution, equation 54. To illustrate the importance of the boundary conditions in determining the form of the solution we can consider the amperometric biosensor system in which the product, P, is amperometrically detected. Note that this case is different from that of the mediated system described in Section 4.1, in which the reduced mediator derived from the product rather than the product itself is detected.

For the amperometric case, the boundary conditions applying to the substrate will be the same as those for the above potentiometric case and the solution for substrate concentration will be given by equation 58. This expression for substrate can be substituted into the differential expression describing the product, equation 51b, to give:

$$\frac{d^2[P]}{dx^2} = \alpha \left(\frac{\cosh(x\sqrt{\alpha})}{\cosh(l\sqrt{\alpha})} \right) [S]_l \tag{63}$$

With the product amperometrically consumed at the electrode surface its concentration there is reduced to zero such that the boundary conditions will be:

$$x = 0 \qquad [P] = 0 \tag{64a}$$

$$x = l \qquad [P] = [P]_l \tag{64b}$$

In this case we are interested not in the surface concentration but the surface flux which is related to the surface concentration gradient. To find this we must solve equation 63, subject to the boundary conditions given by equations 64a and 64b, for the product concentration across the enzyme layer, and then differentiate with respect to distance, x, and substitute $x = 0$. Equation 63 is readily integrated with respect to x to give the general solution:

$$[P] = -\left(\frac{\cosh(x\sqrt{\alpha})}{\cosh(l\sqrt{\alpha})} \right) [S]_l + Ax + B \tag{65}$$

To find the integration constants, A and B, we apply the boundary conditions. First we can substitute $[P] = 0$ at $x = 0$ to give:

$$B = \frac{[S]_l}{\cosh(l\sqrt{\alpha})} \tag{66}$$

and, substituting this into equation 65 gives:

$$[P] = \frac{[S]_l}{\cosh(l\sqrt{\alpha})}[1-\cosh(x\sqrt{\alpha})]+Ax \qquad (67)$$

Now we can substitute $[P] = [P]_l$ and $x = l$ to give the second integration constant:

$$A = \frac{[P]_l}{l}+\frac{[S]_l(\cosh(l\sqrt{\alpha})-1)}{l\cosh(l\sqrt{\alpha})} \qquad (68)$$

and substitute this into equation 67 to give the final solution for the product concentration across the enzyme layer:

$$[P] = \frac{x}{l}[P]_l+\frac{[S]_l}{\cosh(l\sqrt{\alpha})}\left(\frac{x}{l}\{\cosh(l\sqrt{\alpha})-1\}+\{1-\cosh(x\sqrt{\alpha})\}\right) \qquad (69)$$

Note that this differs from the expression derived earlier for the potentiometric case where the sensor does not consume product. It is, however, based upon the same dimensionless parameters; the dimensionless kinetic parameter, $l\sqrt{\alpha}$, and the normalized distance, x/l. Using this expression we can proceed to find the surface flux, by differentiation and substitution of $x = 0$. Also, by substitution of $x = l$ into the differentiated expression, the fluxes in the enzyme layer can be linked with the mass transport flux, in a manner analogous to that illustrated in Section 4.2.3, for the potentiometric case to provide a more complete description of the amperometric system. This is left as an exercise for the interested reader.

5.1.2 Solution of partial differential equations

The way in which partial differential equations arise in the description of coupled transport and chemical kinetics problems relevant to biosensors has been considered in the preceding sections. Generally, the relative complexity of them precludes their solution by simple substitution of an appropriate function, as illustrated in the preceding section for an ordinary differential equation. Two standard methods for the analytical solution of partial differential equations are commonly employed; the Laplace transform and Fourier methods. These can be illustrated by working through the solutions to the time-dependent problems discussed in Section 4.3.

Laplace transform method
The Laplace transform of a function $y = f(t)$ is typically written $L\{y\}$ or $\bar{y}$ and is defined by:

$$\bar{y} = \int_0^\infty y\exp(-st)dt \qquad (70)$$

where s is the variable of the transformed function $\bar{y}$. As we will see, this transformation can be employed to provide from the original partial differential equation an ordinary differential equation describing the transformed function $\bar{y}$, this ordinary differential equation being soluble by normal means to give an algebraic expression for $\bar{y}$. Then, taking the inverse transforms provides the required algebraic expression for original function y. Transforms of functions frequently encountered are to be found in extensive tables (23,24). We can illustrate the application of this method by working through the example given in Section 4.3.2 which is concerned with diffusional mass transport according to Fick's second law, subject to defined initial and boundary conditions. The basic partial differential equation to be solved is:

$$\frac{\partial C}{\partial t} = D\frac{\partial^2 C}{\partial x^2} \tag{71}$$

and is subject to initial and boundary conditions as follows:

$$t = 0, \quad 0 < x < l \qquad C(x,0) = C_i \tag{72a}$$

$$x = l, \quad t > 0 \qquad C(0,t) = 0 \tag{72b}$$

$$x = 0, \quad t > 0 \qquad \left(\frac{dC}{dx}\right) = 0 \tag{72c}$$

We now take the Laplace transforms of both sides of equation 71. The right hand side is quite straightforward since the variable, t, is not directly involved and the transformation is oblivious to other differential operators. We have simply:

$$L\left\{\frac{\partial^2 C}{\partial x^2}\right\} = \frac{\partial^2 \bar{C}}{\partial x^2} \tag{73}$$

For the left hand side, from the definition of the Laplace transform in equation 70 we have:

$$L\left\{\frac{\partial C}{\partial t}\right\} = \int_0^\infty \frac{\partial C}{\partial t}\exp(-st)dt \tag{74}$$

Applying the formula for integration by parts we have:

$$L\left\{\frac{\partial C}{\partial t}\right\} = [C\exp(-st)]_0^\infty + \int_0^\infty sC\exp(-st)dt = -C_i + \bar{C} \tag{75}$$

Note that we have applied the initial condition of equation 72a, defining the initial concentration, C_i, in evaluating this integral expression. Equating these two

Theoretical methods for analysing performance

transforms and making a minor rearrangement, we can now write an ordinary differential equation for the transformed variable as follows:

$$\frac{d^2\bar{C}}{dx^2} = \frac{s\bar{C}}{D} - \frac{C_i}{D} \tag{76}$$

We have already seen the general solution to a differential equation of this type in the preceding section and we may write this in the form:

$$\bar{C} = A\cosh\{x\sqrt{(s/D)}\} + B\sinh\{x\sqrt{(s/D)}\} + \frac{C_i}{s} \tag{77}$$

We can now proceed to evaluate the constants by applying the boundary conditions. First, applying equation 72c we find $B = 0$. Next, applying equation 72b we determine A:

$$A = -\frac{C_i}{s\cosh\{l\sqrt{(s/D)}\}} \tag{78}$$

and hence we have:

$$\bar{C} = -C_i\left(\frac{\cosh\{x\sqrt{(s/D)}\}}{\cosh\{l\sqrt{(s/D)}\}}\right) + \frac{C_i}{s} \tag{79}$$

It is interesting to note the similarity between this solution and that derived in the preceding section and the corresponding similarity in the boundary conditions applied to the diffusion problem. The next task is to take the inverse transform. Methods for performing the inversion are described in standard texts (22) but, typically, the inverse can be found from the available comprehensive tables (23,24) of transforms. From tables we find the inverse of equation 79.

$$C_{(x,t)} = -C_i\left(\frac{4}{\pi}\sum_{n=1}^{\infty}\frac{(-1)^n}{(2n-1)}\exp\left\{-\left(\frac{(2n-1)\pi}{2}\right)^2\frac{Dt}{l^2}\right\}.\cos\left(\frac{(2n-1)n\pi}{2}\frac{x}{l}\right)\right) \tag{80}$$

This is the basic expression for concentration as a function of both distance and time from which the response characteristics of the system can be derived, as discussed in Section 4.3.

Fourier method

The Fourier method is generally less widely used than the Laplace transform method in the solution of time-dependent diffusion problems but has been usefully applied in certain cases. The method assumes separability of the partial differential equation into two or more ordinary differential equations. For our purposes, this means that the function to be found, $C(x,t)$, can be written in terms of the product of two functions, $X(x)$ and $T(t)$, which are entirely separable, according to:

$$C(x,t) = U(x) + X(x)T(t) \qquad (81)$$

where $U(x)$ is a time-independent function, representing the steady-state solution to the problem. What we mean by separability in this context is that the distance dependence takes the same form, described by a time-independent distance function, $X(x)$, whatever the value of time. Similarly the time dependence is described over all distances by the distance-independent function, $T(t)$. Clearly, the application of the method is limited to cases where the various functions describing the time-dependent solution are separable in this manner. To illustrate the application of the method we will consider the time-dependent transmembrane diffusion problem discussed in Section 4.3. As before in the illustration of the Laplace transform method, we are seeking the solution to the diffusion problem described by Fick's second law, according to equation 71, in this case subject to the initial and boundary conditions as follows:

$$t \geq 0, \quad x = 0 \qquad C(0,t) = 0 \qquad (82a)$$

$$t \geq 0, \quad x = l \qquad C(l,t) = C_l \qquad (82b)$$

$$t = 0, \quad 0 < x < l, \qquad C(x,0) = 0 \qquad (82c)$$

We first solve for the steady-state solution, $U(x)$, by setting $\partial C/\partial t = 0$ in equation 71 and solving subject to the boundary conditions given by equations 82b/c. This is straightforward and gives:

$$U(x) = C(0,\infty) + \frac{x}{l}\{C(l,\infty) - C(0,\infty)\} = \frac{x}{l}C_l \qquad (83)$$

We now solve equation 71, in this case for $C(x,t) - U(x) = X(x)T(t)$, rather than $C(x,t)$, subject to initial and boundary conditions relevant to $X(x)T(t)$ derived from those for $C(x,t)$ originally stated in equations 82a/b/c and equation 83:

$$t \geq 0, \quad x = 0 \qquad C(0,t) - U(0) = X(0)T(t) = 0 \qquad (84a)$$

$$t \geq 0, \quad x = l \qquad C(l,t) - U(l) = X(l)T(t) = 0 \qquad (84b)$$

$$t = 0, \quad 0 < x < l, \qquad C(x,0) - U(x) = X(x)T(0) = -\frac{x}{l}C_l \qquad (84c)$$

Theoretical methods for analysing performance

The separable function, $X(x)T(t)$, can be differentiated with respect to x and t to give relationships between its partial and ordinary derivatives as follows:

$$\frac{\partial^2 X(x)T(t)}{\partial x^2} = T(t)\frac{d^2 X(x)}{dx^2} \tag{85a}$$

$$\frac{\partial X(x)T(t)}{\partial t} = X(x)\frac{dT(t)}{dt} \tag{85b}$$

and so, substituting these into equation 71, we have

$$\frac{1}{X(x)}\frac{d^2 X(x)}{dx^2} = \frac{1}{DT(t)}\frac{dT(t)}{dt} \tag{86}$$

With the left hand side dependent upon x only and the right hand side dependent upon t only whilst x and t are completely independent of one another, we deduce that both left hand and right hand sides must be some constant, $\pm\lambda$. We now have two ordinary differential equations to solve:

$$\frac{d^2 X(x)}{dx^2} = \pm\lambda X(x) \tag{87a}$$

$$\frac{dT(t)}{D\,dt} = \pm\lambda T(t) \tag{87b}$$

The general solutions to these ordinary differential equations are quite readily found by consideration and substitution of appropriate functions. However, at this stage we cannot tell whether the constant, λ, is associated with a positive or negative sign and therefore must consider both possibilities. If $\lambda \geq 0$ then equation 87a is of the same form as equation 51a considered in Section 5.1.1. The general solution would be of the same form as equation 54, containing exponential functions. If we now refer to boundary conditions of equations 84a/b, $X(0) = X(l) = 0$, to determine the constants, A and B, we find that the only possible solution is $A = B = 0$ and hence $X(x)T(t) = 0$ throughout the interval $x = 0$ to $x = l$. This solution is trivial and we conclude that λ must take a negative value in order to describe the required concentration function. For this case we note that the trigonometric functions, sine and cosine, will fit equation 87a whilst an exponential function will fit equation 87b. Thus we find the general solutions to equations 87a/b:

$$X(x) = A\cos(\omega x) + B\sin(\omega x) \tag{88a}$$

$$T(t) = C\exp(-\omega^2 Dt) \tag{88b}$$

$$X(x)T(t) = [\overline{A}\cos(\omega x) + \overline{B}\sin(\omega x)]\exp(-\omega^2 Dt) \tag{88c}$$

where $\omega^2 = \lambda$. We must now evaluate the three constants $\bar{A}$, $\bar{B}$ and ω by reference to the three pieces of information provided by the initial and boundary conditions. First, from equation 84a we have $X(0)T(t) = 0$ from which we determine $\bar{A} = 0$ and so equation 88c reduces to

$$X(x)T(t) = \bar{B}\sin(\omega x)\exp(-\omega^2 Dt) \tag{89}$$

Then, from equation 84b, we have $X(l)T(t) = 0$ from which we find $\bar{B}\sin(\omega l) = 0$. If $\bar{B} = 0$ a trivial solution $X(x)T(t) = 0$ is obtained so we conclude that $\sin(\omega l) = 0$ which determines ω:

$$\omega_n = \frac{n\pi}{l} \tag{90}$$

where n is an integer between 0 and ∞. That is, given the nature of the sine function, we have an infinite set of solutions fitting the equation which we are attempting to solve. Substituting these into equation 89, summing each of the possible solutions, gives:

$$X(x)T(t) = \sum_{n=1}^{\infty} \bar{B}_n \sin\left(n\pi \frac{x}{l}\right) \exp\left(-(n\pi)^2 \frac{Dt}{l^2}\right) \tag{91}$$

Finally, we refer to the initial condition defined by equation 84c at $t = 0$ for which $T(0) = 1$ and hence:

$$\sum_{n=1}^{\infty} \bar{B}_n \sin\left(n\pi \frac{x}{l}\right) = -\frac{x}{l}C_l \tag{92}$$

Making use of the orthonormality properties of the sine function, as described elsewhere (22), a value of B_n is obtained:

$$\bar{B}_n = \frac{2(-1)^2}{n\pi} C_l \tag{93}$$

Finally we substitute this value for $\bar{B}_n$ into equation 91 which gives the solution for $X(x)T(t)$ and, referring back to equation 81, the solution for the concentration throughout the layer as a function of time is obtained:

$$C(x,t) = C_l \left[\frac{x}{l} + \sum_{n=1}^{\infty} \frac{2(-1)^2}{n\pi} \sin\left(n\pi \frac{x}{l}\right) \exp\left(-(n\pi)^2 \frac{Dt}{l^2}\right)\right] \tag{94}$$

This function provides the basis for describing the state of the system and its relaxation towards the steady-state condition, as discussed in Section 4.3.

5.2 Numerical solution

In many cases, whilst it may be a relatively straightforward process to write down the differential equation describing a given system, the general complexity of this differential equation will prevent its analytical solution by the methods described above. This is quite commonly the case for the differential equations describing the time dependence of coupled kinetic and mass transport problems of relevance to biosensors and in many cases for those describing the time-independent steady-state behaviour as well. It will then be necessary to employ numerical methods to generate curves relating the variables of interest to one another, these variables typically being the sensor output and analyte concentration. Numerical solution generally involves writing the differential equation in some approximate algebraic form. Values of the various parameters can then be substituted into this algebraic form and the behaviour of the system computed arithmetically. Each solution generated in this way will be specific to the values of the parameters chosen and in order to build up a curve describing the overall behaviour of the system a set of calculations must be made with an appropriate range of values for the relevant variables. There are two numerical methods in common use for the solution of coupled transport and kinetic problems, the finite difference and polynomial approximation methods. These two methods are considered below in turn, following an introduction to the setting up of the basic equation system to be solved.

5.2.1 Definition of the problem: dimensionless variables

We have already seen the use of dimensionless parameters and the way they arise naturally in the analytical solution of various problems. The first task when attempting a numerical solution will typically be to cast the starting equation into a dimensionless form. That is to say a system of dimensionless variables will be defined which represent ratios or relative values of the various parameters describing the system, and these dimensionless variables will be analogous to the dimensionless parameters which arose naturally in the analytical solutions described earlier. To illustrate this we can consider the immobilized enzyme layer described in Section 4.2 for which the time-dependent problem is described by:

$$\frac{d[S]}{dt} = D_s \left(\frac{d^2[S]}{dx^2} \right) - \frac{k_2[E][S]}{K_M+[S]} \tag{95}$$

In this case use will be made of the dimensionless concentration variable, $S^* = [S]/K_M$, the concentration normalized with respect to the Michaelis–Menten constant, and the dimensionless distance variable, $x^* = x/l$, the distance across the enzyme layer normalized with respect to the total layer thickness. We can simply substitute $[S] = S^* K_M$ and $x = x^* l$ into equation 95. Then, defining the dimensionless time parameter, $t^* = Dt/l^2$ and rearranging gives the fully dimensionless form of the equation:

$$\frac{dS^*}{dt^*} = \frac{d^2 S^*}{dx^{*2}} - l^2 \alpha \left(\frac{S^*}{1+S^*} \right) \tag{96}$$

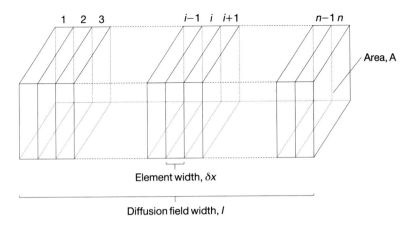

Figure 10. Division of diffusion field into finite volume elements.

where α is as defined earlier by equation 34, where it arose in the analytical solution of the first-order kinetic case and $l^2\alpha$ is a dimensionless parameter representing the relative rates of the various processes involved. The value of using this dimensionless form is as follows. Each particular solution carried out for a given set of values of the relevant dimensionless variable is specific to that case. However, with the dimensionless variables being ratios of the real system variables, each dimensionless solution representing a given ratio represents a multitude of solutions relating to different values of the real system variables corresponding with that ratio. Furthermore, we have reduced the number of variables in our differential equation by half and can therefore solve the mathematical problem in the most efficient way. Thus, equation 96 represents the starting point for the application of numerical methods of solution.

5.2.2 Finite difference methods

Because of their relative simplicity, both conceptually and mathematically, finite difference methods are the numerical methods most commonly applied to the solution of mass transport problems and many accounts of their application to such problems have been given (9,11 and ref therein, 25). Differential equations attempt to give exact descriptions in terms of infinitesimally small changes. Where we cannot solve them we can move to an algebraic representation in terms of very small but finite changes, in our case small changes in substrate concentration, $\delta[S]$, occurring with small changes in time, δt, and distance, δx. If these changes are sufficiently small they will approximate the exact description of the differential equation. So, in order to describe our coupled kinetic and mass transport problem we divide the region of space of interest into an array of n small volume elements, of width δx and cross sectional area, A, orthogonal to this x direction, as illustrated in *Figure 10*. At a given time, t, each element will contain a specified concentration of substrate and

Theoretical methods for analysing performance

we consider the way in which this concentration will alter in a short time, δt, due to chemical reaction and diffusive mass transport. We can consider each volume element in turn and this will provide a set of concentrations relating to the state of the system at the later time, $t+\delta t$. Then, with a series of iterative steps forward in time, the time evolution of the system can be built up. The change in concentration due to chemical reaction rate, $\delta[S]_R$ will be simply the rate of reaction, given by the second term on the right hand side of equation 95, with the appropriate value of substrate concentration for the element under consideration, multiplied by the time, δt, for which that reaction rate is considered to be operative:

$$\delta[S]_R = -\frac{k_2[E][S]}{K_M+[S]}\delta t \tag{97}$$

As we discussed earlier during the development of Fick's laws in Section 3.2, the change in the amount of substrate in a given element due to diffusion, $\delta[S]_D$, is the sum of the fluxes between it and the adjacent elements multiplied by the time for which that flux operates. From Fick's first law, given by equation 13, the flux, j_-, between the ith element, with a concentration, $[S]_i$, and the $(i-1)$th element, with a concentration, $[S]_{i-1}$, will be

$$j_- = -AD\left(\frac{\delta[S]}{\delta x}\right) = -AD\left(\frac{[S]_i-[S]_{i-1}}{\delta x}\right) \tag{98a}$$

A similar expression gives the flux, j_+, between the ith element and the $(i+1)$th element:

$$j_+ = -AD\left(\frac{[S]_{i+1}-[S]_i}{\delta x}\right) \tag{98b}$$

The differences in these fluxes, $j_- - j_+$, multiplied by the time, δt, over which they operate gives the change in amount of substrate in the ith element and, dividing through by the volume, $A\delta x$, of the element, an expression for the concentration change due to diffusion is obtained:

$$\delta[S]_D = D\left(\frac{[S]_{i-1}-2[S]_i+[S]_{i+1}}{\delta x^2}\delta t\right) \tag{99}$$

Adding the contributions to the change in concentration from chemical reaction and diffusion and substituting the dimensionless variables described above gives the basic algebraic expression describing the concentration change in a given volume element:

$$\delta S_i^* = \left\{\left(\frac{S_{i-1}^*-2S_i^*+S_{i+1}^*}{\delta x^{*2}}\right) - l^2\alpha\left(\frac{S_i^*}{1+S_i^*}\right)\right\}\delta t^* \tag{100}$$

This will apply to each of the volume elements apart from those at the boundaries, that is elements 1 and n. To describe these the boundary conditions must be applied. For the example under consideration the boundary condition at $x = 0$ is that the concentration gradient is zero and hence for the first volume element we have $S_0^* = S_1^*$ in equation 100. For the outer boundary relating to the nth element we have a constant, defined concentration, $S_{n+1}^* = S_f^*$.

We have derived the above expression by going back to our physical model and considering the diffusion and chemical reaction processes involved. It is to be hoped that by doing this an understanding of the mathematical process has been gained. However, it is not necessary for us to do this every time. We have already produced a mathematical description of the physical system as the dimensionless differential equation given by equation 96. Without returning to our physical model we could, in an abstract mathematical manner, have written equation 100 directly from equation 96, simply by substituting the general finite difference forms of the relevant derivatives.

Taking equation 100 as a basis, the next step is to build up the time evolution of the system and this is most simply done by what is known as the explicit finite difference method (25) in which the concentrations at time $t^* + \delta t^*$ are calculated explicitly in terms of the concentrations at time t^*. So, with the concentrations on the right hand side of equation 100 designated the old ones at time t^*, $S_{i,\mathrm{OLD}}^*$, and the concentration change being the difference between old and new concentrations, $\delta S_i^* = S_{i,\mathrm{NEW}}^* - S_{i,\mathrm{OLD}}^*$, an expression for the new concentration in terms of the old is obtained:

$$S_{i,\mathrm{NEW}}^* = S_{i,\mathrm{OLD}}^* + \left[\left(\frac{S_{i-1,\mathrm{OLD}}^* - 2S_{i,\mathrm{OLD}}^* - S_{i+1}^*}{\delta x^{*2}}\right) - l^2\alpha\left(\frac{S_{i,\mathrm{OLD}}^*}{1 + S_{i,\mathrm{OLD}}^*}\right)\right]\delta t^* \quad (101)$$

We start with the initial condition that the concentration is everywhere zero within the membrane and constant at its bulk value at the outer boundary and apply equation 101 to each volume element. Values of each of the dimensionless variables are first defined and then, by an iterative arithmetic procedure, comprising calculation of new concentrations followed by substitution of these as the next set of old concentrations, the time-dependent response of the system can be built up.

It is worth noting briefly that the explicit method has a draw back in that it is unstable if insufficiently small time steps are taken. That is, the inaccuracy in the set of concentrations calculated gets progressively worse with each time iteration as these inaccuracies compound themselves until the errors become exceedingly large and the solution no longer follows the equations upon which it is based. Taking sufficiently small time steps to ensure stability can require an excessive amount of computer time and it is therefore often found more convenient to employ the implicit finite difference method (26). In this method the concentrations on the right hand side of equation 100 are designated as the new ones. The finite difference expression for each element will therefore contain three unknowns, that is the new concentration in the $(i-1)$th, the ith and the $(i+1)$th elements, except for the elements at the boundaries where one of these concentrations will be defined. In this case it will not be possible to calculate each concentration explicitly. However, there are n equations of this type and a total of n unknowns such that these equations can be solved simultaneously to give each of the required concentrations. It is also worth noting that, if it is the steady-state solution rather than the time dependence that is of interest, putting $\delta S^*/\delta t^* = 0$ in equation 100 again gives a set of n simultaneous equations which can be solved (27) to give the n unknown concentrations.

5.2.3 Polynomial approximation methods

Polynomial approximation methods for the solution of differential equations (28) have been applied to the solution of diffusion problems (29) and are, in many cases, more efficient in computational terms than are finite difference methods. However, they are more advanced mathematically and typically require more mathematical skill to implement. Finite difference methods have therefore often been preferred. It will be worthwhile to consider briefly the principle behind the polynomial approximation method which is as follows. The function for which the solution is sought is assumed to be in the form of a polynomial and in our example we define substrate concentration as a function of distance:

$$S^* = a_0 + a_1 x + a_2 x^2 + \ldots + a_i x^i + \ldots + a_n x^n \tag{102}$$

where $a_0, a_1, \ldots a_i, \ldots a_n$ are a set of coefficients which we wish to determine to provide a solution for the concentration function. This expression can be differentiated with respect to x and the second derivative is, for example, given by

$$\frac{d^2 S^*}{dx^2} = 2a_2 + 6a_3 x \ldots i(i-1) a_i x^{i-2} + \ldots n(n-1) a_n x^{n-2} \tag{103}$$

What this has now produced is an algebraic expression for the second derivative. Substitution of this expression can then be employed to reduce an ordinary differential equation to an algebraic expression and a partial differential equation to an ordinary differential equation. Methods for the solution of these resulting equations, that is for evaluation of the polynomial coefficients, are described (27–30) elsewhere.

References

1. Berzofsky, J. A. and Berkower, I. J. (1984). In *Fundamental Immunology*. Paul, W. E. (ed.), Raven Press, New York, p. 595.
2. DeLisi, C. (1976). *Antigen Antibody Interactions: Lecture Notes on Biomathematics* Vol. 8. Levin, S. (ed.), Springer-Verlag, Berlin.
3. Karush, F. (1978). In *Immunoglobulins*. Litmann, G. W. and Good, R. A. (eds), Plenum, New York, P. 85.
4. Oscik, J. (1982). *Adsorption*. Eliis Horwood Ltd, Chichester.
5. Eddowes, M. J. (1987/88). *Biosensors*, **3**, 1.
6. Jencks, W. P. (1980). *Mol. Biol. Biochem. Biophys.*, **32**, 3.

7. Laidler, K. J. and Bunting, P. S. (1973). *The Chemical Kinetics of Enzyme Action.* 2nd edition, Clarendon Press, Oxford.
8. Carr, P. W. and Bowers, L. D. (1980). *Immobilized Enzymes in Analytical and Clinical Chemistry: Chemical Analysis Vol. 56.* Elving, P. J. and Winefordner, J. D. (eds), John Wiley and Sons, New York.
9. Bard, A. J. and Faulkner, L. R. (1980). *Electrochemical Methods: Fundamentals and Applications.* John Wiley and Sons, New York.
10. Bird, R. B., Stewart, W. E., and Lightfood, E. N. (1960). *Transport Phenomena.* John Wiley and Sons, New York.
11. Crank, J. (1975). *The Mathematics of Diffusion.* 2nd edition, Clarendon Press, Oxford.
12. Levich, V. G. (1962). *Physicochemical Hydrodynamics.* Prentice-Hall, Englewood Cliffs, NJ.
13. Albery, W. J. and Hitchman, M. (1971). *Ring-Disc Electrodes.* Clarendon Press, Oxford.
14. Compton, R. G. and Unwin, P. R. (1986). *J. Electroanal. Chem.,* **205**, 1.
15. Albery, W. J. and Brett, C. M. A. (1983). *J. Electroanal. Chem.,* **148**, 210.
16. Chin, D.-T. and Tsang, C.-H. (1978). *J. Electrochem Soc.,* **125**, 1461.
17. Albery, W. J. and Craston, D. H. (1987). In *Biosensors: Fundamentals and Applications.* Turner, A. F. P., Karube, I., and Wilson, G. S. (eds), Oxford University Press, Oxford, p. 180.
18. Albery, W. J. and Bartlett, P. N. (1985). *J. Electroanal. Chem.,* **194**, 211.
19. Eddowes, M. J. (1987). *Sensors and Actuators,* **11**, 265.
20. Badley, R. A., Drake, R. A. L., Shanks, I. A., Smith, A. M., and Stephenson, P. R. (1987). *Philos. Trans. R. Soc. Lond.,* **316**, 143.
21. Stephenson, G. (1973). *Mathematical Methods for Science Students.* Longman, London.
22. Stephenson, G. (1970). *An Introduction to Partial Differential Equations for Science Students.* Longman, London.
23. Roberts, G. E. and Kaufman, H. (1966). *Tables of Laplace Transforms.* Saunders, Philadelphia.
24. Abramowitz, M. and Stegun, I. (1964). *Handbook of Mathematical Functions.* National Bureau of Standards, Washington.
25. Feldberg, S. W. (1969). In *Electroanalytical Chemistry.* Bard, A. J. (ed.), Vol. 3, p. 199.
26. Smith, G. D. (1969). *Numerical Solutions of Partial Differential Equations.* Oxford University Press, Oxford.
27. Eddowes, M. J. (1983). *J. Electroanal. Chem.,* **159**, 1.
28. Villadsen, J. and Michelsen, M. L. (1978). *Solution of Differential Equation Models by Polynomial Approximation.* Prentice-Hall, Englewood Cliffs, NJ.
29. Whiting, L. F. and Carr, P. W. (1977). *J. Electroanal. Chem.,* **81**, 1.
30. Rees, D. C. (1984). *Bull. Math. Biol.,* **46**, 229.

Appendix

Suppliers of reagents, membranes, sensors, and associated equipment

Biological materials

An excellent source of information on suppliers of enzymes, antibodies, and other biological materials is *Linscott's Directory of Immunological and Biological Reagents*. This is regularly updated and also has supplements published during the life of each edition. It is available from Linscott's Directory, 40 Glen Drive, Mill Valley, CA 94941, USA, and in Europe from Linscott's Directory, PO Box 55, East Grinstead, Sussex RH19, 3YL, UK.

Sources of enzymes specifically mentioned in the text are:

Sigma Chemical Company, PO Box 14508, St Louis, MO 63178, USA, and **Sigma Chemical Co Ltd**, Fancy Rd, Poole, Dorset BH17 7TG.
Kyowa Hakko Kogyo Co, Tokyo, Japan.

Membranes

Amicon, Danvers, MA, USA and **Amicon Ltd**, Upper Mill, Stonehouse, Glos GL10 2BJ, UK.
Centre du Cuir, Lyon, France.
FMC Inc., Princeton, NJ, USA.
Garlock Plastomer Products, Newton, PA, USA.
Millipore Ltd, Elkton, MD, USA and **Millipore Europe Ltd**, 11–15 Peterborough Rd, Harrow, Middx HA1 2YH, UK.
Nucleopore Corp, 7035 Commerce Circle, Pleasanton, CA 94556, USA.
In the UK from **Sterilin Ltd**, Sterilin House, Clockhouse Lane, Feltham, Middx TW14 8QS, UK.
Pall, Glen Cove, NY 11542, USA, and **Pall Europe Ltd**, Europa House, Havant, Portsmouth PO1 3PD, UK.
Thermedics Inc., Woburn, MA, USA.
Viscase Corp, Chicago, IL, USA.
In the UK from **Medicell International Ltd**, 239 Liverpool Road, London N1 1LX, UK.

Sensors

ABB Kent, Oldends Lane, Stonehouse, Glos GL10 3TA, UK.
Oxford Electrodes, 62 Main Road, Hoo, Rochester, Kent ME3 9AB, UK.
Radiometer-Tacussel, Villeurbanne, France.
In the UK from **V.A.Howe**, 12–14 St Anne's Crescent, London SW18 2LS, UK.
SERES, Aix-en-Provence, France.
SGI, Toulouse, France.
Veco, Victory Engineering Co, Springfield, NJ, USA.
Yellow Springs Instruments Inc., Yellow Springs, OH 45387, USA.
In the UK from **Clandon Scientific Ltd**, Lysons Avenue, Ash Vale, Aldershot, Hants GU12 5QR, UK.

Potentiostats and other instrumentation

Aber Instruments, Aberystwyth Science Park, Cefn Llan, Aberystwyth SY23 3AH, UK.
Analogic Corp, Wakefield, MA, USA.
Bioanalytical Systems Inc, West Layfayette, IN, USA.
In the UK from **Anachem Ltd**, Charles St, Luton, Beds, LU2 0EB, UK.
EG&G Princeton Applied Research, PO Box 2565, Princeton, NJ 08540, USA and **EG&G Instruments**, Doncastle House, Doncastle, Bracknell RG12 4PG, UK.
Hewlett-Packard Ltd, Miller House, The Ring, Bracknell, Reading RG12 1XN, UK.
Knauer Wissenschaftlicher, West Berlin, FRG.
In the UK from **Owens Polyscience Ltd**, 34 Chester Rd., Macclesfield, Cheshire SK11 8DG, UK.
Malthus Instruments Ltd. The Manor, Manor Royal, Crawley, W. Sussex RH10 2PY, UK.
Orbit Biotechnology Ltd, PO Box 26, Deiniol Rd, Bangor, Gwynedd LL57 2UG, UK.
Oxford Electrodes, 62 Main Road, Hoo, Rochester, Kent ME3 9AB, UK.
Radiometer-Tacussel, Villeurbanne, France.
In the UK from **V.A.Howe**, 12–14 St Anne's Crescent, London SW18 2LS, UK.
Sensitor AB, Box 376, S-58102, Lingkoping, Sweden.
Solartron Instruments Ltd, 124 Victoria Rd, Farnborough, Hants GU14 7PW, UK.
Sycopel Scientific Ltd, Station Road, East Boldon, Tyne and Wear NE36 0EB, UK.
Thermometric Co, Jarfalla, Sweden.
TOA Electronics, Tokyo, Japan.
In the UK from **Centronic Sales Ltd**, King Henry's Drive, New Addington, Croydon CR9 0BG, UK.

Ancillary equipment

Churchill Instruments, Riverside Way, Uxbridge, Middx, UK.
Gilson, Villiers-le-Bel, France.
In the UK from **Anachem Ltd**, Charles St, Luton, Beds, LU2 0EB, UK.
Gould Bryans Instruments Ltd, Willow Lane, Mitcham, Surrey CR4 4UL, UK.
Heto, Klintehoj Vaenge 3, 2460 Birkeroed, Denmark.
LKB, Bromma, Sweden.
Mitsumi Scientific Industry, Tokyo, Japan.

Other materials

Biotrol Pharma, Paris, France.
Buehler, PO Box 150, Binns Close, Coventry CV4 9XJ, UK.
Engis Ltd, Parkwood Trading Estate, Maidstone, Kent ME15 9NJ, UK.
Norton Performance Plastics, Chesterton Works, Loomer Rd, Newcastle, Staffs ST5 7HR, UK.
Pharmacia-LKB, Uppsala, Sweden.
Rheodyne, Cotali, CA 94928, USA.
Rohm-Pharma, Weiterstadt, FRG.
RS Components, PO Box 99, Corby, Northants, NN17 9RS, UK.
Touzart et Matignon, Vitry, France.
Wacker-Chemitronic, Burghausen, FRG.

INDEX

admittance 125
agar 159
AIROF 139
alcohol dehydrogenase 83
alcohol oxidase 200
alkaline phosphatase
  properties 103
ammonia
  sensitive MOS 175,186
D-amino acid oxidase 81
L-amino acid oxidase 81
p-aminophenol 106,110
p-aminophenyl phosphate 106
ampicillin 205
anodic iridium oxide film electrode, see AIROF
antibodies 97,212
antigens 97,212
ascorbate oxidase 197
ATP 30,203
azurin 43

batch measurement 162
p-benzoquinone 198
bile acids 86,87
4,4'-bipyridyl 35
biomass 148
bis(4-pyridyl)disulphide 38
boundary conditions 222
Brownian motion 220
BSA 5,6
butyrylcholinesterase 145

caffeine 107
calcium alginate 202
calibration 14,15,79,85,116,123
capacitance 127
capillary fill cell 246
casein 161
catalase 2,198−200
catalytic currents 23
cellulose acetate 3,10,77,157
  p-benzoquinone activation 6

cholesterol 32,199
cholesterol dehydrogenase 32
cholesterol oxidase 199
  reaction with ferrocenes 33
choline oxidase 81,199
chronoamperometry 103
coenzyme recycling 203
concentration polarization 87,219,241
conductance 127
conducting organic salt electrodes
  analysis of performance 89
conducting organic salts
  enzyme adsorption 79
  reaction with glucose oxidase 75
  structures 48
  stable potential range 73
conductivity 128
controlled pore glass, see CPG
collagen 4
  acyl azide activation 5
Cottrell equation 223
counter electrode
  platinum gauze 58
CPG 181,196
cream 144
creatine kinase 29
creatinine 168,187
creatinine iminohydrolase 187
p-cresolmethylhydroxylase 43
crosslinking 5
cyanide 206
cyclic voltammetry 70,101
cytochrome c 35,40
cytochome oxidase 40

Debye equation 129
dielectric loss 129
dielectric spectrum 132
diffusion 87
diffusion coefficient 21
diffusion layer 224
dimensionless variables 236,259
dispersion 130
double layer capacitance 135
double reciprocal plot 214,218
dual sensor 12,81,146

electrochemical cells 10,25,64
  rotating disk 65
electrodes
  basal plane 37
  CO_2 160
  edge plane 37
  fuel cell 160
  galvanic 159
  graphite 24
  interdigitated 147
  NH_3 160
  paste 67
  pH 160
  polarographic 159
  pyrolytic graphite 26
  rotating disk 226
enthalpies of enzyme reactions 191
entrapment 3
enzyme layer 3,233
enzyme loading factor 238,250
epitope 98
Escherichia coli 205
ethanol 200
Eupergit C, *see* oxirane acrylic beads
explicit finite differences 262

ferrocenes
  reaction with flavocytochrome b_2 42
  reaction with glucose oxidase 22,28
  solubilization with detergents 123
  theophylline conjugate 123
FET 172
Fick's laws of diffusion 220
field effect transistor, *see* FET
finite difference 261
flavocytochrome b_2 41
flow through sensor 10,164,183
foodstuffs 148
Fourier method 254
Frequency Response Analyser 142
fructose 161

galactose 198
galactose oxidase 198
β-galactosidase 203
Gluconobacter oxydans 202
glucose 198
  interfering substances 9
  microbial sensor 161
glucose oxidase 3
  periodate oxidation 119
  properties 116
  reaction with conducting organic salts 77
  reaction with ferrocenes 24,30

glucose-6-phosphate dehydrogenase 203
glutamate dehydrogenase 187
glutaraldehyde 4
glycerol 203

Hanes plot 90
hapten 98
heavy metals 206
hexokinase 30,198,203
hydrogen dehydrogenase 179,185

immobilization
  cellulose acetate 6
  collagen 5
  CPG 181
  graphite 22
  oxirane acrylic beads 181,188
  polyamide (nylon) 7,113
immunoassays
  competitive 99
  displacement 99
  heterogeneous 99
  homogeneous 99
  sandwich 100
immunogens 98
immunoglobulins 97
impedance 128
Impedance Analyser 142
implicit finite difference 261
insulin 207
invertase 197
ion sensitive field effect transistor *see* ISFET
ISFET 171,173

β-lactamase 181,201
L-lactate 199,203
lactate dehydrogenase 41,203
lactate oxidase 199,203
lactate-2-monoxygenase 199
lactose 203
Langmuir adsorption isotherm 216
Laplace transform 245,254
Lineweaver-Burke Plot 218
lipoprotein lipase 199

magneto-inductance 150
mass transfer rate constant 88
mass transport 219
mediators
  hexacyanoferrate (III) 19,118

organic dyes 17,118
membrane
  entrapment 3
  protective 2,8
metal oxide semiconductor, *see* MOS
Michaelis constant 218
microelectrodes 45,80
microsensor 11
MOS 171

NAD$^+$ 185
  dextran modified 83
NADH 32,83,185,203
Nafion 9
naphthyl phosphate 106
needle sensor 11
Nernst equation 18,101,173
Newton's law of viscosity 224
nitrocellulose 157
non-polarizable 134
nylon
  activated 7,113
nystatin 168

ordinary differential equations 248,254
oxalate 200
oxirane acrylic beads 181,196

partial differential equations 255
penicillin 184,201
permittivity 129
peroxidase
  reaction with ferrocene 33
phase angle 125
phase-sensitive detector 142
phenyl phosphate 106
phenytoin 109,111,114
phospholipids 199
phospholipase D 199
photodetectors 161
plastocyanin 36
platinum black electrode 137
polarizable 136
polarization 136
polyacrylamide 158
polyamide
  activated (Biodyne) 7
polycarbonate 4,8,23
polytetrafluoroethylene membrane 182
polyurethane 9,10
product inhibition 89
potentiostat 13,61
Pseudomonas sp. S-17 166

2-pyridylmethylenehydrazinecarbothioamide 43
pyruvate kinase 203

rate limiting step 87
reactance 127
reference electrode
  calomel 57
  potentials 61
  silver/silver chloride 60
relaxation time 243
resistance 127
response time 15
reversible electrochemistry 23
rhodanese 206
rotational relaxation time 132

Saccharomyces cerevisiae 151,167
Scatchard plot 215
Sepharose 4B 207
Sepharose CL-6B 196
signal to noise ratio 15
skin 151
Spectrum Analyser 144
steady state assumption 217
Stokes-Einstein equation 131
sucrose 198
surface modifiers 35
susceptance 128

theophylline 107,114
  ferrocene conjugate 121
thermal enzyme probe 192
thin metal film oxide semiconductor, *see* TMOS
TMOS 171
tresyl chloride 207
Trichosporon cutaneum 165
triglycerides 199
two electrode and three electrode
  systems 13,18,57

urea 146,187,192,198
urease 146,187,198,206

vitamin C 198

Warburg impedance 135
Wheatstone bridge 141

xanthine oxidase 81